Fluids and Electrolytes
with Clinical Applications

WILEY PAPERBACK NURSING SERIES

FLUIDS AND ELECTROLYTES
WITH
CLINICAL APPLICATIONS

A Programmed Approach

Joyce LeFever Kee

Assistant Professor, Medical-Surgical Nursing
College of Nursing, University of Delaware
Newark, Delaware

John Wiley & Sons, Inc.   New York / London / Sydney / Toronto

Dedicated to my husband, Edward,
for his understanding and helpfulness,
and to my children, Eric, Katherine, and Wanda.

Library of Congress Catalogue Card Number: 70-154229

ISBN 0-471-46400-7

Printed in the United States of America

10 9 8 7 6 5 4 3

Wiley Paperback Nursing Series

This unique series provides the nursing profession with current and timely information. The five volumes do not follow a single theme. Some volumes deal with topics that are particularly interesting to undergraduate and graduate college students, some deal with matters that appeal to nursing educators, and others consider problems that confront practicing nurses and persons in the nursing service.

Many of these books are appropriate for use in both graduate and undergraduate courses. The contributors have selected their subjects with care. Each author possesses the interest, background, and experience necessary to deal comprehensively with the chosen subject.

Ever-expanding general knowledge and the rapid advances in the medical and health fields make it imperative that nurses have readily available materials that are relevant to their needs. I believe this series accomplishes this purpose and that it is an important contribution to the nursing profession.

Nurses, teachers, and students alike should find these books helpful and stimulating.

MILDRED MONTAG
Teachers' College,
Columbia University

Preface

At least ten years ago, nurses did not need to understand fluid and electrolyte imbalance, but today, in the era of automation and specialization, assessment of fluid and electrolyte imbalance by nurses is of utmost importance. Medical science is one of the most rapidly growing fields in the country. The nurse must be knowledgeable of medical advances and be cognizant of the changes which occur with patients as a result of these advances. She can no longer perform all her nursing functions because she was told to do so, but, instead, she must have the knowledge of why—the rationale. Also, the nurse must know more than "something is wrong with my patient," for many times some immediate care could be given while awaiting the physician.

This programmed instruction on *Fluids and Electrolytes with Clinical Applications* enables the participant to work at her own pace while learning the principles, concepts, and applications of fluids and electrolytes which are presented in this text.

Learning that takes place is efficient, pleasant, and permanent, for the individual proceeds through the programmed chapters by: participating in a large number of small steps, actively responding to these steps, and receiving immediate confirmation to her answers.

Programmed instruction has rapidly gained wide acceptance in both general and specialized education. The material in this text is geared to three levels within the nursing profession. First, for the beginning students who have had some background in the biological sciences or an anatomy and physiology course. Second, for the students who have had a sufficient backgound in the biological sciences, chemistry, and physics and need the clinical conditions and situations that are found within this text. Third, for helping the graduate nurse review and increase her knowledge of fluid and electrolyte changes for assessing patients' needs and for the improvement of patient care.

This method of learning will aid the instructor in using the class time to better advantage. It is felt that this book will enable students to apply their learned knowledge to other clinical situations in their clinical practicum.

There are 60 diagrams and tables, each followed by frames that help the participants gain a deeper understanding of the material. The first three chapters give general information on fluid and electrolyte balance and imbalance and each chapter has clinical considerations included. The fourth chapter is on parenteral therapy and contains clinical considerations, nursing interventions, and rationale. The fifth chapter contains four main clinical conditions: dehydration, water intoxication, edema, and shock which are programmed in detail to enable the student or graduate nurse to be cognizant of these conditions when they confront her. Also included in this chapter are clinical considerations, clinical examples (an individual in marked dehydration and one in congestive heart failure) and clinical management of these conditions. The sixth chapter includes clinical situations which can cause severe fluid and electrolyte imbalance. They are: gastrointestinal surgery, cirrhosis of the liver and ascites, renal failure with peritoneal dialysis, burns, and diabetic acidosis. Also, these chapters include physiological factors, clinical considerations, clinical examples, and clinical management. Since the nurse frequently encounters these situations, it is necessary that she understand the fluid and electrolyte changes that occur which could result in the life or death of the individual. A glossary is included with words and terms used throughout the text. This should assist the student who has had only some preparation in the biological sciences.

Many "reviews" occur thoughout the chapters to cover the subject matter discussed and reinforce the learning. Also, there are summary questions at the end of each chapter for the individual to review the knowledge gained from his participation. In the Appendix, there are five clinical situations presented for the nurse's assessment. It represents a guide for nursing interventions.

After completing this book, the students and/or graduate nurses should understand more fully the effects of fluid and electrolyte balance and imbalance on the body in many conditions and clinical situations.

I would like to express my appreciation to the students, who, for 3 years, used this programmed instruction book in varied forms. From testing their acquired knowledge and application after completion of this text, I was able to make the necessary changes. I would also like to extend my appreciation to Elizabeth Kee, for her secretarial assistance in preparation of this book, Eric Gregory, a medical student at the University of Delaware, for his illustrated diagrams found in this book, Mrs. Carolyn Behr, an instructor in programmed instruction, at the University of Delaware, and to Mrs. Irene Brownstone for her most helpful assistance with this programmed text revision.

January 1971   Joyce LeFever Kee
                                                  New Castle, Delaware

Instructions

TO THE INSTRUCTOR

Much class time is frequently spent on reviewing material or presenting new material which could easily be given through programmed instruction. This enables the instructor to minimize the time spent in lecture on fluids and electrolytes and, thus, she can devote more time to clinical conferences and seminar classes as they relate to clinical situations with fluid and electrolyte imbalance.

You may find it helpful to cover the material in this book by one of three ways: (1) assigning the students a chapter at a time; (2) assigning the students the first three chapters to be completed by a certain date and the last three by a later date; or (3) assigning the students a given length of time to complete the entire text.

TO THE STUDENTS

Many students feel the subject of fluids and electrolytes is most difficult and hard to comprehend. This programmed book will provide you with important data on fluids and electrolytes from various aspects and you will find this material is not so difficult to understand and retain.

By taking the easy steps provided in this book, you will proceed through the chapters more quickly than you would expect. This book is written on a self-instruction basis so that you may proceed at your own speed. Each step is a learning process. Greater learning occurs if you either complete a chapter or spend a minimum of two (2) hours at one sitting. Never end the study period without at least completing all the frames related to a single topic.

It would be helpful to begin your next study session with the final frames from the previous material, for it will enable you to check your retention of the material which was presented. The "reviews" throughout the chapters give immediate reinforcement of the data learned. A glossary is included to assist you with words and terms used in the text.

Students, study the diagrams before proceeding to the frames. If you make mistakes in the program, you need not be concerned so long as you rectify the mistakes. This book should increase your knowledge and understanding of fluids and electrolytes and be a great asset to applying this knowledge in your clinical practicum.

USE OF TEXT

In programmed instruction instead of simply reading information, you actively participate in the programmed exercises that have been set up for you. The information is broken down into organized units called frames.

Each frame either presents some new information or reviews material presented earlier. Each frame imparts information, asks a question, and requires one or more responses. You will be asked to make the response by writing a word or phrase, completing a chart, or selecting several correct answers from alternatives. You may write your responses in the book or on a separate piece of paper.

Use the answer shield, or a similar card to cover the correct response (just below the dashed line) before you start reading the first frame on the page. After you have made your response, compare your answer with the correct response(s). An asterisk (*) on an answer line indicates a multiple-word answer. If you make an error in responding to a frame and do not understand why you made the error, then refer to previous frames. If you do understand the reason for your error, advance to the next frame.

SAMPLE FRAME

Water is composed of two elements, hydrogen and oxygen.

Name the two elements which make up water.

_____ and _____ .

_____

hydrogen and oxygen

or

The two elements responsible in the formation of water are:

( ) a. helium and oxygen
( ) b. hydrogen and oxygen
( ) c. hydrogen and sulphur

_____

a. —
b. X
c. —

REVIEW INSTRUCTION

The reviews found within the chapters are to reinforce learning from the material presented. The participant should follow the same procedure for the review frames as presented in the general instructions; that is, covering the answer with a card.

Contents

Chapter 6 (continued)

Fluids and Electrolytes
with Clinical Applications

Chapter 1

Body Fluid and Its Function

BEHAVIORAL OBJECTIVES

Upon completion of this chapter, the student will be able to:

Differentiate between the percentage of water found in the average adult body, newborn infant, and the embryo.

Name the three compartments of the body where body water is found.

Name the two classifications of body fluid and their percentages.

Define homeostasis and explain its state in maintaining body fluid equilibrium.

Explain how the body loses and maintains body fluid.

Define osmotic pressure, semipermeable and selectively-permeable membranes, osmol and osmolality, and relate the effects of each in the passage of body fluid.

Differentiate between isotonic, hypotonic, and hypertonic solutions and explain the effects of these solutions on body cells.

Define milliequivalent and milligram and explain their symbols and significance in the body.

Define Starling's Law of Capillaries.

Explain the four measurable factors that determine the flow of fluid between the vessels and tissues and their effects on the exchange of fluid.

Explain the pressure gradient and the significance in colloid osmotic and hydrostatic pressure gradients.

Relate fluid changes that occur to patients in your clinical area.

INTRODUCTION

The human body is a complex machine that contains hundreds of bones and the most sophisticated systems of any structure on earth. Yet, the substance that is basic to the very existence of the body is the simplest substance known—water. In fact, it makes up almost two-thirds of an adult's body weight.

The body is not static—it is alive, and solid particles within its framework are able to move into and out of cells and systems, and even into and out of the body, only because there is water.

The basis of all fluids is water, and as long as the quantity and composition of body fluids are within the normal range, we just take it for granted and enjoy being healthy. But, if the water content of the body for some reason departs from this range, the whole delicate balance is disrupted, and disease can find an easy target.

In this chapter, distribution of body fluid, fluid compartments of the body, physics terminology, fluid pressures, and clinical considerations are discussed.

1

The greatest single constituent of the body is water, which represents about 60% of the total body weight in the average adult. In the early human embryo, 97% of body weight is water, in a newborn infant 77%.

Label the following drawings with the proper percentage of water to body weight.

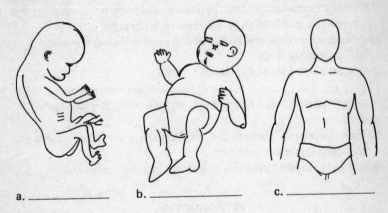

a. _____   b. _____   c. _____

- - - - - - - - - - - - - - - -

a. 97%
b. 77%
c. 60%

2

In the average adult, the proportion of water to _____ is 60%.

- - - - - - - - - - - - - - -

body weight

_____

3

Which has the highest percentage of water in relation to body weight?

_____

Which has the lowest? _____

- - - - - - - - - - - - - - -

embryo, adult

_____

4

Can you think of any reason why the early human embryo and the infant have a higher proportion of water to body weight?

- - - - - - - - - - - - - - -

I asked what you thought. Many people think the extra water in infants acts as a protective mechanism. Since infants have larger body surface in relation their weight, extra water acts as a cushion against injury.

_____

5

Because fat is essentially free of water, the leaner the individual, the greater the proportion of water in total body weight.

Who has more water as body weight—a fat person or a lean one?

- - - - - - - - - - - - - - -

lean person

6

This body water is distributed among three types of "compartments":
cells, blood vessels, and tissue spaces between blood vessels and cells,
which are separated by membranes.

Label the three compartments where body water is found.

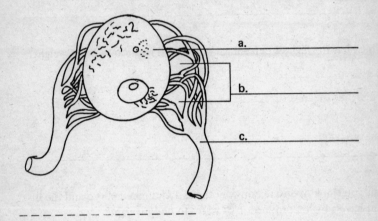

a. _____

b. _____

c. _____

— — — — — — — — — — —

a. cell
b. tissue space
c. blood vessel

7

The term for the water in each type of "compartment" is as follows:

1. In the cell—intracellular water or cellular water.
2. In the blood vessels—intravascular water.
3. In tissue spaces between blood vessels and cells—interstitial water.

Label the complete diagram with the proper terms for body water in the three compartments.

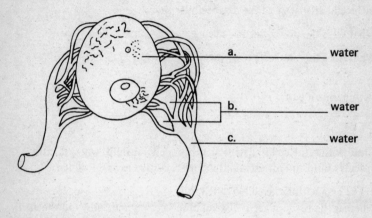

a. _____ water

b. _____ water

c. _____ water

- - - - - - - - - - - - - - -

a. intracellular water
b. interstitial water
c. intravascular water

8

There are three prefixes that will be used frequently in this text:

inter-   between
intra-   within
extra-   outside of

The two classes of body water or body fluid are: intracellular fluid and extracellular fluid. The corresponding areas or "compartments" are the intracellular space and the extracellular space.

What does the prefix "intra" mean? _____

What does the prefix "extra" mean? _____

_ _ _ _ _ _ _ _ _ _ _ _ _ _ _ _

within, outside of

9

Fluid within the cell is classified as intracellular fluid, whereas intra-vascular fluid and interstitial fluid are classified as extracellular.

   The area within the cell is called the _____ space, whereas the tissue spaces between blood vessels and cells, and the area

within blood vessels are known as the _____ space.

_ _ _ _ _ _ _ _ _ _ _ _ _ _ _ _

intracellular, extracellular

10

Label the three "compartments" of body water and the two classes of body water.

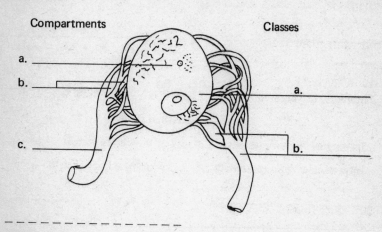

Compartments   Classes

a. _____

b. _____

                           a. _____

c. _____

                           b. _____

– – – – – – – – – – – – – – – –

Compartments
a. cell
b. tissue space
c. blood vessel

Classes
a. cellular or intracellular
b. extracellular

11

Approximately two-thirds of the body fluid is contained in the intracellular space.

We have already said that the total water in the adult body is _____%; therefore, intracellular water must represent _____ % of the total body weight, and extracellular water _____ %.

– – – – – – – – – – – – – – – –

60%, 40%, 20%

12

If one-fourth of the extracellular fluid is intravascular water, then three-fourths is _____.

_ _ _ _ _ _ _ _ _ _ _ _ _ _ _

interstitial fluid

13

Therefore, extracellular fluid represents _____ % of body weight; of which _____ is interstitial fluid.

   Interstitial fluid represents _____ % of total body weight.

   Intravascular fluid represents _____ % of total body weight.

_ _ _ _ _ _ _ _ _ _ _ _ _ _ _

20%, three fourths, 15%, 5%

14

Since we already have learned that the percentage of body fluid varies with age and "fatness," then the proportion of intracellular and extracellular fluid in a fat person would be (greater/lesser) to his body weight.

_ _ _ _ _ _ _ _ _ _ _ _ _ _ _

lesser

Review

Q. 1

The greatest single constituent of the body is _____.

_ _ _ _ _ _ _ _ _ _ _ _ _ _ _

water

Q. 2

In an adult, this constituent represents _____ % of total body weight, in a human embryo _____ % of body weight, and in a newborn infant _____ % of body weight.

– – – – – – – – – – – – – –

60%, 97%, 77%

Q. 3

Extra water in infants acts as a protective mechanism. Since infants have larger body surface in relation to their weight, extra water acts as a

*_____ .

– – – – – – – – – – – – – –

cushion against injury

Q. 4

A lean person has (more/less) body water than the fat person.

– – – – – – – – – – – – – –

more

Q. 5

Complete the following chart.

Compartments	Names of water in compartments	Classes of body fluid	Percentage of body fluid in an adult
Cells			
Blood vessels			
Tissue spaces			

– – – – – – – – – – – – – – –

Compartments
Intracellular water, Intravascular water, Interstitial water

Classes
Intracellular or cellular, Extracellular, Extracellular

Percentage
Cells 40%, Blood vessels 5%, Tissue spaces 15%

15

Homeostasis is a term you will be using. Homeostasis means the state of equilibrium of the internal environment. Concerning body fluids, homeostasis is the maintaining of equilibrium or stability in relation to the physical and chemical properties of body fluid.

In a few words define the term homeostasis. _____

Explain the relationship of homeostasis to body fluid. _____

_____

– – – – – – – – – – – – – – –

state of equilibrium, maintains equilibrium of the physical and chemical properties of body fluid

16

The body normally maintains a state of equilibrium between the intake and loss of water.

If the body loses or gains water, it acts rapidly to compensate for this

deficit or excess so that _____ will be maintained.

— — — — — — — — — — — — — — —

homeostasis, or equilibrium

17

When body water is insufficient, the urine volume diminishes, and the

individual becomes thirsty. Therefore, he might_____
to make up the deficiency.

— — — — — — — — — — — — — — —

drink more water

18

Do you think man can go longer without food or without water? _____

Why do you think man cannot go without water? _____

— — — — — — — — — — — — — — —

If you said he can go longer without food, this is correct. Again water is the greatest single constituent of the body.

Because water is contained in the various compartments of our body and is needed to carry nutrients, elements, and waste to and from our body tissues.

19

If we drink an excessive amount of water, the urinary output would increase.

If you did not drink any fluids or if the body lost water, then the urinary volume would:

(   )   a. increase
(   )   b. decrease

If there were an excess of water in the body, then the urinary volume would:

(   )   a. increase
(   )   b. decrease

— — — — — — — — — — — — — — —

b. decrease

a. increase

Intake
Liquid 1200 ml
Food 1000 ml
Oxidation of Food 300 ml

Output

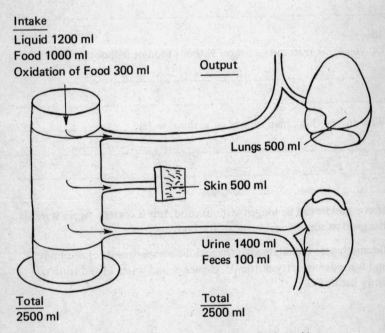

Lungs 500 ml

Skin 500 ml

Urine 1400 ml
Feces 100 ml

Total
2500 ml

Total
2500 ml

Diagram 1. Normal pattern of water intake and loss.

20

Refer to Diagram 1. The normal sources of body water are _____ ,

_____ , and _____ .

- - - - - - - - - - - - - -

liquid, food, and oxidation of food

21

Refer to Diagram 1. The avenues for daily water loss are _____ ,

_____ , _____ , and _____ .

- - - - - - - - - - - - - -

lungs, skin, urine, and feces

22

If your water intake amounted to 2500 ml for the day and your water output was 2500 ml, then your body has maintained a state of

_____ of body fluid.

- - - - - - - - - - - - - -

equilibrium or homeostasis

23

Would you think that the rate of water loss and gain is different in summer and winter, or the same all year round? _____ .

- - - - - - - - - - - - - -

In the summer when the atmospheric temperature is high, water loss, via skin and lungs, increases.

24

Evaporation of water from the skin, as we perspire, is a protective mechanism against overheating the body.

Can you explain how evaporation of water acts as a protective

mechanism?_____.

– – – – – – – – – – – – – – – –

It acts as a cooling system, keeping the body at a normal temperature.

25

Do you know the meaning of the term diffusion?

Diffusion means the movement of each molecule along its own pathway irrespective of all other molecules. A short definition would be the spreading of molecules through a membrane in various directions.

Which would be considered diffusion—molecules going in the same direction or in various directions irrespective of other molecules?

– – – – – – – – – – – – – – – –

various directions

26

Body water loss by diffusion through the skin which is independent of sweat gland activity is called insensible perspiration.

When sweat gland activity occurs and water appears on the skin, this is called sensible perspiration.

In a relative comfortable temperature would insensible perspiration or

sensible perspiration occur? _____ . Why? _____

_____.

– – – – – – – – – – – – – – – –

Insensible.
There is not enough heat to cause sweat gland activity, so only the normal loss occurs, through insensible perspiration, with water diffusing through the skin and evaporating quickly.

27

Give the meanings of the following words:
Insensible Perspiration.
Sensible Perspiration.

– – – – – – – – – – – – – – – –

Insensible—Water loss by diffusion through the skin.
Sensible—Water on the skin due to sweat gland activity.

28

The volume of body water is primarily regulated by the kidneys. When water loss increases, e.g., through perspiration or diarrhea, the kidneys will conserve water by (increasing/decreasing) the urinary output.

– – – – – – – – – – – – – – –

decreasing

29

Some more definitions to know:
Membrane. A layer of tissue covering a surface or organ, or separating spaces.
Osmosis. The passage of a solvent through a partition separating solutions of different strength.
Solvent. A liquid with a substance in solution.
Solute. A substance dissolved in a solution.
Permeability. A capability of a substance, molecule, or ion to diffuse through a membrane.

Early literature described membranes of body cells as semipermeable. Today semipermeable refers to artificial membranes, i.e., cellophane membrane, frequently described in the process of osmosis. Selectively permeable membrane refers to the human membrane.
   Differentiate between:
Selectively permeable membrane: _____ .
Semipermeable membrane: _____ .

– – – – – – – – – – – – – – –

human membrane, artificial membrane

30

Do you recall the difference between a solvent and a solute? Explain.

Solvent. _____

_____.

Solute. _____

_____.

— — — — — — — — — — — — — — —

Solvent—A liquid with a substance in solution.
Solute—A substance dissolved in solution.

31

In an effort to establish equilibrium, water in the body moves from a less concentrated solution (fewer solute particles per unit of solvent) to a more concentrated solution (more solute particles per unit of solvent) through a _____membrane.

— — — — — — — — — — — — — —

selectively permeable or human

32

Osmotic pressure is the pressure or force that develops when two solutions of different strengths or concentrations are separated by a selectively permeable membrane.

In osmosis, to establish equilibrium, water would move from the (less/more) concentrated solution into the (less/more) concentrated solution.

The force that draws water across a selectively permeable membrane is called _____.

— — — — — — — — — — — — — —

less, more, osmotic pressure

33

In what direction does water flow? _____

_____ . Why? _____

– – – – – – – – – – – – – –

From the lesser to the greater concentration. Because more solute particles have a "pulling" effect.

34

Do you recall the meaning of permeable? If not, return to Frame 29.

A membrane is considered impermeable if an ion, substance, or molecule cannot diffuse freely across it.

Certain substances do not diffuse freely across the human membrane, so this membrane is considered _____ to that substance.

– – – – – – – – – – – – – – –

impermeable

35

The kidney is influenced to excrete or conserve water by ADH, which is the antidiuretic hormone. This hormone is excreted from the posterior pituitary gland, also called the posterior hypophysis.

Do you know where the pituitary gland, also called the hypophysis,

is? _____ .

– – – – – – – – – – – – – –

Yes? Fine.
No? Consult a physiology book.

36

The antidiuretic hormone, or ADH, increases the permeability of the cells of the kidney tubules to water, thus allowing more water to be reabsorbed. With a lack of this hormone, what would occur? _____

_____ .

– – – – – – – – – – – – – –

an increased excretion of water from the kidney tubules

37

Two terms that need to be defined:

<u>Serum</u>. Consists of plasma minus fibrogen. It is obtained after coagulation of blood.

<u>Plasma</u>. Contains blood minus the blood cells. It is composed mainly of water.

Frequently these two terms are interchangeably used. Also the term blood plasma means plasma and blood serum means serum.

   Serum and plasma are both found in what type of fluid?

_ _ _ _ _ _ _ _ _ _ _ _ _ _

intravascular fluid or extracellular fluid

38

The posterior pituitary gland is influenced by the solute concentration of the plasma. If there is an increase solute in the plasma, then the gland will release the hormone ADH which will hold water in the body.

Explain how? _____ .

   For what reason should there be more water? _____ .

_ _ _ _ _ _ _ _ _ _ _ _ _ _

It will absorb water from the kidney tubules.
To dilute the solute.

39

A small increase of solute concentration in the plasma, above the normal amount, would be sufficient to stimulate the posterior gland in releasing

_____.

  What two things would occur if there was less solute concentration in the plasma?
1.
2.

– – – – – – – – – – – – – – – – –

ADH
1. ADH would not be released.
2. More water would be excreted from the body.

40

If you drank a lot of fluids, what would happen to the solute concentration of your blood plasma—would it become diluted or more concentrated?

  Then what would the posterior pituitary do?

– – – – – – – – – – – – – –

diluted, Would not release ADH

41

When the solute concentration increases, the thirst mechanism is stimulated, and the individual ingests water.

  From the above statement how can homeostasis be maintained? _____

_____.

– – – – – – – – – – – – – –

By drinking water or other liquids.

42

An osmol is a unit of osmotic pressure. The osmotic effects are expressed in terms of osmolality.
A milliosmol (mOs) is 1/1000th of an osmol and will determine the osmotic activity.

Three terms to know:

Ion. A particle carrying a positive or negative charge. (A further explanation of ion will be found in Chapter 2.)
Dissociation. Separation, i.e., a compound separating into many particles.
Molar. 1 Gram molecular weight of a substance.

Which is larger, the osmol or the milliosmol? _____ .

Do you know the relationship between ion and dissociation? _____

_____ .

– – – – – – – – – – – – – – – –

osmol, Ion is a particle from dissociation of a compound.

43

According to Frame 42, 1 milliosmol is 1/1000th of an osmol. Then 1 osmol would equal _____ milliosmols.

Milliosmols will determine the _____ activity of a solution.

– – – – – – – – – – – – – –

1000, osmotic

44

An osmol is determined by dividing the molar concentration of a solute by the number of ions formed in dissociation.

The basic unit used to express the force exerted by the concentration

of solute or dissolved particles is a(n) _____ .

The osmotic effect of a solute concentration is expressed as

_____ , a property that depends on the number of osmols or milliosmols.

― ― ― ― ― ― ― ― ― ― ― ― ―

osmol, osmolality

45

Here are three more prefixes which will be used frequently in this text:

    hypo-   less than
    iso-   equal
    hyper-   excessive

The osmolality of a solution is isotonic if it has the same solute concentration within the solution as does plasma, hypotonic if it has a lower concentration, and hypertonic if it has a higher concentration.

A solution is classified by comparing its osmolality with that of

_____ .

Plasma is considered to be a(n) (isotonic/hypotonic/hypertonic) fluid.

― ― ― ― ― ― ― ― ― ― ― ― ―

plasma, isotonic

46

Match the following types of solutions with their solute concentration.

____1. Isotonic   a. Higher solute concentration than plasma
____2. Hypotonic   b. Same solute concentration as plasma
____3. Hypertonic   c. Lower solute concentration than plasma

― ― ― ― ― ― ― ― ― ― ― ― ―

1. b, 2. c, 3. a

47

The osmolality of blood plasma is 290 mOs. Parenteral solutions or solutions for intravenous use having 290 mOs with either +50 mOs or −50 mOs would be considered an isotonic solution.

A solution having less than 240 mOs is considered _____ ,

and a solution having more than 340 mOs is considered _____ .

– – – – – – – – – – – – – – –

hypotonic, hypertonic

48

The following chart contains a list of milliosmol values. Classify them as isotonic, hypotonic, or hypertonic.

Milliosmol Values	Type of Osmolality
220 mOs	_____
75 mOs	_____
350 mOs	_____
310 mOs	_____
560 mOs	_____

– – – – – – – – – – – – – – –

hypotonic, hypotonic, hypertonic, isotonic, hypertonic

49

If two solutions have the same osmol values, their osmotic pressures will be the same.

Intracellular fluid in the cells is isotonic fluid. Distilled water has a lower osmotic pressure than the cells, thus, it is considered to be

_____ fluid.

– – – – – – – – – – – – – – –

hypotonic

50

Extracellular hypertonic fluid has a greater osmotic pressure than the cell, thus, intracellular water moves out of the cells and into the extracellular hypertonic fluid by the process of _____ .

With the cells losing water, what would happen to their form and size? _____ .

- - - - - - - - - - - - - - - -

osmosis, Cells would shrink and become small in size.

51

If the extracellular fluid was hypotonic, what would happen to the cells? _____ .

- - - - - - - - - - - - - - - -

Cells would swell and enlarge in size.

52

A flask of 5% dextrose in water is 250 mOs and a flask of 0.9% sodium chloride or normal saline is 310 mOs having somewhat the same osmotic pressure as _____ .

These solutions would be (isotonic/hypotonic/hypertonic) fluid.

- - - - - - - - - - - - - - - -

plasma, isotonic

53

The sum of 5% dextrose in normal saline would equal _____ mOs.

This solution would be a(n) _____ solution.

- - - - - - - - - - - - - - - -

560 mOs, hypertonic

54

Name some isotonic solutions:

1.

2.

— — — — — — — — — — — — — — —

1. 5% dextrose
2. 0.9% sodium chloride or normal saline
3. plasma

55

In studying blood serum chemistry alterations and concentrations, one is concerned with how much the ions or chemical particles weigh, which are measured in milligrams percent or mg % (mg % means the same as mg/100 ml), or with the number of electrical charged ions, which is measured in milliequivalents per liter (1000 ml) or mEq/L.

The term milliequivalent involves the chemical activity of elements while milliosmol involves the _____ activity of solution.

How do milligrams and milliequivalents differ?

— — — — — — — — — — — — — — —

osmotic
Milligrams—the weight of ions.
Milliequivalents—the chemical activity of ions

56

Actually milliequivalents is a better method of measuring the concentration of ions in the blood serum than milligrams.

Milligrams measure the _____ of ions and give no information concerning the number of ions or the number of electrical charges of ions.

— — — — — — — — — — — — — — —

weight

57

If you were having a party and want to invite equal numbers of boys and girls, which would be more accurate—inviting 1500 pounds of girls

and 1500 pounds of boys or inviting 15 girls and 15 boys?_____

Why? _____

_____ .

– – – – – – – – – – – – – – –

15 girls and 15 boys. You could have an unequal number of boys and girls for not every child weighs 100 pounds.

58

From the example in Frame 57, which would be more accurate in determining the blood serum chemistry of chemical particles or ions in the

body—milliequivalents or milligrams? _____ .

   You will find both measurements used in this book and in your clinical settings in determining changes in our blood serum chemistry. However, when referring to ions, milliequivalents will be used in this book.

– – – – – – – – – – – – – – –

milliequivalents

59

Milligrams are reported in terms of mg % or mg/100 ml, and milliequivalents are reported in terms of mEq/L or mEq/1000 ml.

Complete the following chart filling the uncompleted symbols and
words.

Symbols		Meanings
5 m ____ /L.	=	_____ per liter
145 _____ /1000 ____	=	_____ per 1000 ml
82 ____ %	=	_____ per %
110 _____ /100 ml	=	_____ per 100 ml

– – – – – – – – – – – – – –

mEq/L = milliequivalent
mEq/1000 ml = milliequivalent
mg % = milligrams
mg/100 ml = milligrams

Review

Q. 1
Explain the meaning of homeostasis. _____
_____.

– – – – – – – – – – – – – – –

It maintains equilibrium of the physical and chemical properties of
body fluid.

Q. 2
If there is excess intake of water, the urine output (increases/
decreases).

– – – – – – – – – – – – – –

increases

Q. 3

The normal sources of water intake are:

(   ) a. liquids
(   ) b. food
(   ) c. lungs
(   ) d. oxidation of food
(   ) e. skin

_ _ _ _ _ _ _ _ _ _ _ _ _ _ _

a. X
b. X
c. —
d. X
e. —

Q. 4

The normal avenues for daily water loss are:

(   ) a. lungs
(   ) b. skin
(   ) c. food
(   ) d. urine
(   ) e. feces

_ _ _ _ _ _ _ _ _ _ _ _ _ _ _

a. X
b. X
c. —
d. X
e. X

Q. 5

Water evaporating quickly from the skin surface independent of sweat
gland activity is called (sensible/insensible) perspiration.

_ _ _ _ _ _ _ _ _ _ _ _ _ _ _

insensible

Q. 6
What is the function of ADH? _____

_____ .

‒ ‒ ‒ ‒ ‒ ‒ ‒ ‒ ‒ ‒ ‒ ‒ ‒ ‒ ‒

Hormone causing kidney tubules to reabsorb water.

Q. 7
When there is an increase in solute concentration of blood serum, the posterior pituitary gland, or posterior hypophysis, produces (more/less) ADH. This influences the kidneys to (excrete/conserve) water.

‒ ‒ ‒ ‒ ‒ ‒ ‒ ‒ ‒ ‒ ‒ ‒ ‒ ‒ ‒

more, conserve

Q. 8
If the solute concentration of the serum decreases, ADH (will/will not) be released. This influences the kidneys to (excrete/conserve) water.

‒ ‒ ‒ ‒ ‒ ‒ ‒ ‒ ‒ ‒ ‒ ‒ ‒ ‒ ‒

will not, excrete

Q. 9
The force that is dependent on the concentration of solute particles is

known as *_____ .

‒ ‒ ‒ ‒ ‒ ‒ ‒ ‒ ‒ ‒ ‒ ‒ ‒ ‒ ‒

osmotic pressure

Q. 10
An osmol is a unit of _____ . An osmol is deter-

mined by dividing the _____ concentration of a _____ by the number of ions formed in dissociation.

‒ ‒ ‒ ‒ ‒ ‒ ‒ ‒ ‒ ‒ ‒ ‒ ‒ ‒ ‒

osmolality or osmotic pressure, molar, solute

Q. 11

The osmotic pressure of a solution is known as the _____
of the solution.

– – – – – – – – – – – – – – –

osmolality

Q. 12

A solution with osmolality similar to that of plasma is considered to be
(isotonic/hypotonic).

– – – – – – – – – – – – – – –

isotonic

Q. 13

When one is studying blood serum chemistry alterations, he is inter-

ested in the *_____ of ions and the _____ .

They would be expressed as _____ and _____ .

– – – – – – – – – – – – – – –

chemical activity, weight, mEq/L and mg %.

60

E. H. Starling states that equilibrium exists at the capillary membrane when the fluid leaving circulation equals exactly the amount of fluid returning to circulation.

Factors regulating the flow of blood constituents between the interstitial and intravascular compartments are known as Starling's Law of Capillaries. There are four measurable factors that determine the flow of fluid between the two compartments.

Where does equilibrium exist when fluid leaves and returns to circulation?_____.

Starling's Law of Capillaries is concerned with what two compartments? _____ and _____.

— — — — — — — — — — — — — — —

capillary membrane—emphasis on the equilibrium is on the fluid flow at the capillary membrane.
intravascular and interstitial

61

Two terms to define:

Colloid. A nondiffusible substance, a solute suspended in solution.
Hydrostatic. A state of equilibrium of fluid pressures.

The measurable factors incorporated in Starling's Law of Capillaries are the colloid osmotic pressure and the hydrostatic pressure of both the blood in the intravascular compartment and the tissues surrounding the interstitial compartment.

The measurable factors that involve the intravascular compartment are the colloid osmotic pressure and the hydrostatic pressure of the (blood/tissue).

The measurable factors involving the interstitial compartment are the colloid osmotic pressure and the hydrostatic pressure in the

_____.

— — — — — — — — — — — — — — —

blood, tissues

62
What are the two measurable factors that determine the flow of fluid between the intravascular compartment or vessels and interstitial compartment or tissues?

1.
2.

– – – – – – – – – – – – – – – –

1. colloid osmotic pressure
2. hydrostatic pressure

63
Colloid means *_____; so therefore, colloid osmotic pressure would be the amount of pressure exerted from

*_____ .

   Hydrostatic is the *_____ of fluid; so therefore, hydrostatic pressure would be the amount of pressure at

_____ of fluid.

– – – – – – – – – – – – – – – –

nondiffusible substances, nondiffusible substances
state of equilibrium, equilibrium

64
The colloid osmotic pressure and the hydrostatic pressure of the blood

and tissues move fluid through the_____membrane.

– – – – – – – – – – – – – – –

capillary

65

Do you know the meanings of arterioles and venules? If not:

<u>Arterioles</u> are minute arteries that lead into a capillary bed.
<u>Venules</u> are minute veins that lead from the capillary.

Which would be larger, the arteriole or the artery? _____.

The venule or the vein? _____.

— — — — — — — — — — — — — —

artery, vein

66

Fluid exchange occurs only across the walls of capillaries and not across the walls of arterioles or venules. Therefore, fluid moves into the interstitial space at the arteriolar end of the capillary and out of the interstitial space into the capillary at the _____.

— — — — — — — — — — — — —

venular end

67

Gases move from an area of higher concentration to an area of lower concentration. This is the opposite of the movement of fluid in osmosis, when fluid moves from the (less/more) concentrated solution into the (less/more) concentrated solution.

  Since oxygen is in greater concentration in the capillaries, it moves into the interstitial fluid. Carbon dioxide passes in the opposite direction, from the higher concentration in the interstitial fluid to the

* _____ in the capillaries.

— — — — — — — — — — — — — —

lower concentration

68

In the capillaries, oxygen is in greater concentration and, therefore,

moves *_____ .

Carbon dioxide is in higher concentration in the interstitial fluid and,

therefore, moves *_____ .

_ _ _ _ _ _ _ _ _ _ _ _ _ _ _

into the interstitial fluid, into the capillaries

69

The capillary endothelium or capillary membrane acts as a selectively
permeable membrane by permitting free passage of crystalloids. Crystalloids are diffusible substances, and will dissolve in solution. They are
noncolloid substances.

Albumin, protein, and gelatin are colloids of what type of substance?

_____.

The amount of osmotic pressure that develops at a membrane depends

mainly on the concentration of *_____ .

An example of a nondiffusible substance is _____ .

_ _ _ _ _ _ _ _ _ _ _ _ _ _ _

nondiffusible, nondiffusible substances or colloids,
albumin, protein, or gelatin

70

There is a current question as to whether proteins are colloids or crystalloids, however, for our purposes, we shall consider protein a colloid. According to Gibbs-Donnan theory of membrane equilibrium, the osmotic pressure of an ionized colloidal system such as plasma is also due in part to the unequal distribution of diffusible ions. Therefore, 30% of the osmotic pressure developed in normal plasma is due to the unequal distribution of sodium and chloride ions.

Osmotic pressure of an ionized colloidal system is due in part to the

*_____ of diffusible ions.

_ _ _ _ _ _ _ _ _ _ _ _ _ _ _ _

unequal distribution.

71

Blood contains blood cells and plasma. The red blood cell is normally bathed in plasma. Plasma and red blood cells have an equal quantity of nondiffusible solutes, so would there or would there not be an osmotic

pressure at the cell membrane? _____ .

If the red blood cells were bathed in pure water, the presence of large quantities of nondiffusible substances inside the cell would cause

*_____ .

_ _ _ _ _ _ _ _ _ _ _ _ _ _ _

there would not be, the cells to swell and maybe burst

72

Fluid flows only when there is a difference in pressure at the two ends of the system. This difference in pressure between two points is known as the pressure gradient.

If the pressure at one end was 32 and at the other end was 26, then

the pressure gradient would be _____ .

_ _ _ _ _ _ _ _ _ _ _ _ _ _

6

73

The plasma in the capillaries have hydrostatic pressure and colloid osmotic pressure. The tissue fluids have hydrostatic pressure and colloid osmotic pressure.

The difference of pressure between the plasma colloid osmotic pressure and the tissue colloid osmotic pressure is known as the

*_____ .

The difference of pressure between the plasma hydrostatic pressure and the tissue hydrostatic pressure is known as the

*_____ .

It is this difference in pressure which makes the fluid flow.

– – – – – – – – – – – – – – – –

colloid osmotic pressure gradient, hydrostatic pressure gradient

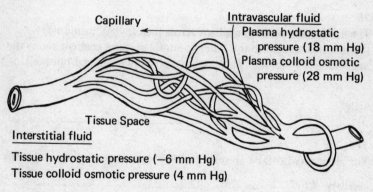

Diagram 2. The pressures in the intravascular and interstitial fluids.

74

Refer to Diagram 2. The plasma colloid osmotic pressure is 28 mm Hg (millimeters of mercury) and the tissue colloid osmotic pressure is 4 mm Hg.

The colloid osmotic pressure gradient would be _____.

- - - - - - - - - - - - - - -

24 mm Hg

75

Refer to Diagram 2. The hydrostatic fluid pressure is 18 mm Hg in the capillary and the hydrostatic tissue pressure is −6 mm Hg; therefore,

the hydrostatic pressure gradient is _____.

- - - - - - - - - - - - - - -

24 mm Hg

76

The hydrostatic pressure gradient across the capillary membrane (24 mm Hg) is equal to the colloid osmotic pressure gradient across the membrane (24 mm Hg). Thus, the two pressure are (equal/unequal).

- - - - - - - - - - - - - -

equal

77

The plasma hydrostatic pressure gradient tends to move fluid out of the

capillary. Why? _____.
Refer to Diagram 2 if reply is unknown.

The colloid osmotic pressure gradient tends to move fluid into the

capillary. Why? _____.

- - - - - - - - - - - - -

The plasma hydrostatic pressure is higher than the tissue.
Plasma pressure is higher than tissue.

78

The balance between the two forces would keep blood volume constant for circulation. In this way fluid would not accumulate in the intra-vascular or the interstitial compartments.

Without the colloid osmotic forces, fluid (would/would not) be lost

from circulation. Do you know the reason why? _____

_____ .

The blood volume would then be (sufficient/insufficient) to maintain circulation.

– – – – – – – – – – – – – – – –

would
Fluid would stay in the tissues causing accumulation and tissue swelling.
insufficient

79

Name the man who formulated the Law of Capillaries and define this law in your own words.

Name: _____

Law: _____

_____

_____ .

– – – – – – – – – – – – – – –

Starling
Plasma and tissue colloid osmotic and hydrostatic pressures regulate the flow of blood constituents between the interstitial and intravascular compartments.

Clinical Consideration

80

There are several diseases that affect the plasma colloid osmotic pressure due to the loss of serum protein.

Several terms need defining:

Protein. A nitrogenous compound, essential to all living organisms.
   Plasma protein relates to albumin, globulin, and fibrinogen.
   Serum protein relates to albumin and globulin.
Serum albumin. Simple protein. Contains the main protein in the blood—about 50%.
Serum globulin. A group of simple protein.

Patients with diagnoses of kidney and liver diseases or malnutrition will lose serum protein.

   What are the two groups of simple proteins found in the blood serum?

_____ and _____.

_ _ _ _ _ _ _ _ _ _ _ _ _ _ _

albumin and globulin

_____

81

The main function of serum albumin is to maintain the colloid osmotic pressure of blood.

   Without colloid osmotic pressure, what would happen to the fluid in

the tissues? _____ .

_ _ _ _ _ _ _ _ _ _ _ _ _ _ _

Fluid would accumulate and swelling would occur. This is known as edema.

_____

82

Serum globulin is not fully understood but one of its functions is to assist in maintaining colloid osmotic pressure of the blood.

The globulin molecule is larger than the albumin molecule, gram for gram, and it is less effective in maintaining osmotic pressure. Serum albumin will leak out of the capillaries in what types of diseases?

_____ .

The serum globulin will be retained trying to compensate for the loss of albumin. As stated, globulin is not as effective as albumin in maintaining osmotic pressure, so what would happen to the colloid osmotic

pressure in the capillaries? _____ .
And to the fluid in the interstitial spaces or compartments?

_____ .

– – – – – – – – – – – – – – –

liver and kidney diseases and malnutrition
lower osmotic pressure
accumulation and swelling would occur

83

Frequently the physician orders an A/G ratio when the probable diagnosis is a kidney or liver disease.

From the previous frames, what do you think is the meaning of A/G?

_____ .

A shift in the A/G ratio will aid the physician in diagnosing the patient's illness.

– – – – – – – – – – – – – – –

albumin and globulin—very good

84

What do you think are some of the nursing responsibilities when caring for patients with abnormal serum albumin and serum globulin?

1.

2.

_ _ _ _ _ _ _ _ _ _ _ _ _ _ _ _

I asked what you thought—possible answers:

1. report abnormal serum findings immediately.
2. observe and report physical findings of swelling or edema.
3. keep an accurate record of fluid intake and output

85

Edema or swelling occurs when there is fluid retention, and dehydration occurs with excess fluid removal.

If the osmolality of intravascular fluid is greater than the osmolality of intracellular fluid, would the cells lose or gain water? _____.

Would edema or dehydration occur to the cells? _____.

_ _ _ _ _ _ _ _ _ _ _ _ _ _ _ _

lose, dehydration

86

With any venous obstruction or vein obstruction, there is an increased venous hydrostatic pressure. This in turn inhibits the fluid moving out of the tissues, causing the tissues to *_____

_____ .

_ _ _ _ _ _ _ _ _ _ _ _ _ _ _

accumulate fluid and causing swelling

87

Normal circulation of blood is dependent upon differences in hydrostatic pressure in the arteries, capillaries, and veins.

Increased hydrostatic pressure in the veins would *_____

_____ .

— — — — — — — — — — — — — —

prevent circulation and cause swelling of the tissues

88

Do you know the meaning of lymph and lymphatic system. If not:

Lymph. Alkaline fluid. Similar to plasma except that the protein content is lower in the lymph vessels.
Lymphatic System. It involves the conveyance of lymph from the
tissues to the blood.

Additional information can be secured from a physiology book.

The lymphatic system has the role of removing excess proteins and
fluid from the interstitial tissue, thus returning protein to the circulation.

Interference with the lymphatic circulation will cause a blockage of
lymph flow and will inhibit the return of tissue protein to the circulation. With an increased tissue lymph fluid and an increased tissue pro-

tein, what would occur to the tissues? _____ .

— — — — — — — — — — — — — —

fluid retention or swelling

Review

Q. 1

Factors regulating the flow of body constituents between the interstitial and intravascular compartments are stated by *_____
_____.

— — — — — — — — — — — — — — —

Starling's Law of Capillaries

Q. 2

The difference in pressure between two points in a fluid is known as the *_____.

— — — — — — — — — — — — — — —

pressure gradient

Q. 3

Pressure gradients are responsible for the exchange of fluid between the capillaries and the _____.

— — — — — — — — — — — — — — —

tissues

Q. 4

The amount of colloid osmotic pressure that develops depends on the concentration of nondiffusible substances such as_____.

— — — — — — — — — — — — — — —

protein or albumin

Q. 5

The direction of the movement of fluid depends on the resultants of the opposing forces. The hydrostatic pressure is greater than the colloid osmotic pressure at the arterial end of the capillary, thus the fluid moves

out of the _____ into the _____ .
    The osmotic pressure is greater than the hydrostatic pressure at the venous end of the capillary, thus the fluid moves out of the

_____and reenters the _____ .

– – – – – – – – – – – – – – –

capillary into the surrounding tissues, tissues and reenters the capillaries

Q. 6

If an individual loses plasma protein, (edema/dehydration) frequently occurs.

– – – – – – – – – – – – – – –

edema

Q. 7

An increased venous hydrostatic pressure will (inhibit/increase) venous return.

– – – – – – – – – – – – – – –

inhibit

SUMMARY QUESTIONS

The questions in this summary will test your comprehensive ability and knowledge of the material found in in Chapter 1. If the answer is unknown, refer to the section in the chapter for further understanding and clarification.

1. What is the percentage of water to the body weight in an early human embryo, a newborn infant, and an average adult?

2. What are the two classes of body fluid? Explain their compartments and the percentage of water to body weight (of an adult) found in their compartments.
3. Define the following terms:

   a. Homeostasis
   b. Sensible and insensible perspiration
   c. ADH
   d. Osmotic pressure
   e. Osmol
   f. Osmolality
   g. Hypotonic, isotonic, and hypertonic solutions
   h. Pressure gradient
   i. Colloid osmotic pressure
   j. mEq/L and mg %
   k. Diffusion
4. Explain how the posterior pituitary gland or the posterior hypophysis plays an important role in homeostasis.
5. Give examples of isotonic solutions.
6. What is Starling's Law of Capillaries? State and/or explain the four measurable factors that determine the flow of fluid.
7. What is the chief constituent responsible for maintaining fluid in the intravascular space or compartment? The capillary membrane acts as a selectively permeable membrane for the free passage of

   _____.

8. What occurs with the body fluid of individuals having kidney or liver disease, or malnutrition, when there is a loss of serum protein?

References

Abbott Laboratories, *Fluid and Electrolytes*, North Chicago, Ill., 1968, p. 4–13.

Best, Charles H. and Norman B. Taylor, *The Physiological Basis of Medical Practice*, 8th ed., Baltimore, Md.: Williams and Wilkins Co., 1966, pp. 595–608, p. 1270.

Bland, John W., *Clinical Metabolism of Body Water and Electrolytes*, Philadelphia, Pa.: W. B. Saunders Co., 1963, pp. 24–26.

Burgess, Richard, "Fluids and Electrolytes," *American Journal of Nursing*, LXV (October, 1965), pp. 90–93.

Flitter, Hessel H., *An Introduction to Physics in Nursing*, 4th ed., St. Louis, Mo.: C. V. Mosby Co., 1962, pp. 98–102.

Garb, Solomon, *Laboratory Tests in Common Use*, 4th ed., New York: Springer Publishing Co., Inc., 1966.

Guyton, Arthur C., *Function of the Human Body*, 2nd ed., Philadelphia, Pa.: W. B. Saunders Co., 1964, pp. 43–60.

Guyton, Arthur C., *Textbook of Medical Physiology*, 3rd ed., Philadelphia, Pa.: W. B. Saunders Co., pp. 424–427, pp. 433–444.

Jacob, Stanley W. and Clarice A. Francone, *Structure and Function in Man*, Philadelphia, Pa.: W. B. Saunders Co., 1965, pp. 454–461.

Kimber, Diana C., *et al.*, *Anatomy and Physiology*, 14th ed., New York: The Macmillan Co., 1961, pp. 576–587.

Kleiner, Israel S. and James M. Orten, *Biochemistry*, 7th ed., St. Louis, Mo.: C. V. Mosby Co., 1966, pp. 30–32.

Langley, L. L., *Outline of Physiology*, New York: The Blakiston Division, McGraw-Hill Book Co., 1961, pp. 261–263.

Nardi, George L. and George D. Zuidema, eds., *Surgery*, 2nd ed., Boston, Mass.: Little Brown and Co., 1965, pp. 137–140.

Pace, Donald, *et al.*, *Physiology and Anatomy*, New York: Thomas Y. Crowell Co., 1965, pp. 21–23.

Taber, Clarence W., *Taber's Cyclopedic Medical Dictionary*, 10th ed., Philadelphia, Pa.: F. A. Davis Co., 1965.

Chapter 2

Electrolytes and Their Influence on the Body

BEHAVIORAL OBJECTIVES

Upon completion of this chapter, the student will be able to:

Explain the relationship of nonelectrolytes, electrolytes, and ions.
Name the principal cation and anion of the extracellular and intracellular fluids.
Give the normal ranges of serum potassium, sodium, calcium, magnesium, and chloride.
Explain the various clinical conditions causing potassium, sodium, calcium, magnesium, and chloride deficits or excesses.
Give the signs and symptoms of hypo-hyperkalemia, hypo-hypernatremia, hypo-hypercalcemia, hypo-hypermagnesemia, and hypo-hyperchloremia.
Relate the electrolyte imbalances to clinical conditions.
List the classes of food which are rich in potassium, sodium, calcium, magnesium, and chloride.

INTRODUCTION

Chemical compounds may behave in one of two ways when placed in solution. In one way, their molecules may remain intact as in urea, dextrose, and creatinine in the body fluid. These molecules do not produce an electrical charge and are considered nonelectrolytes.

The other type of compound develops a tiny electrical charge when dissolved in water. The compound breaks up into separate particles known as ions and this process is referred to as ionization. These compounds are known as electrolytes. Some electrolytes develop a positive charge when placed in water, while others develop a negative charge.

The chemical composition of sea water and human body fluid are very similar. The principal cations of sea water are: sodium, potassium, magnesium, and calcium; and so it is with the body fluid. The sea water contains as principal anions: chloride, phosphate, and sulfate, and so does body fluid.

In this chapter, the five electrolytes—potassium, sodium, calcium, magnesium, and chloride—are discussed as related to the human body needs, the physical effect of having an excess or a deficit, prevalent diseases causing or resulting from electrolyte imbalance, and foods that are rich in these electrolytes. Clinical considerations are discussed with each electrolyte except chloride. The clinical consideration for chloride will be presented in Chapter 3 as it relates to acid-base balance.

1

Electrolytes are compounds which when placed in solution will conduct an electric current.

Pure water does not conduct electricity, but if a pinch of salt, which contains sodium and chloride, is dropped into it, what do you think

would happen to the water? _____.
Refer to the Introduction if needed to answer.

— — — — — — — — — — — — — —

Salt would produce an electrical charge. Sodium and chloride are ions, found in sea water and in our body.

2

Ions are dissociated particles of electrolyte that carry either a positive charge called cation or negative charge called anion.

Dissociated particles of electrolyte are called _____ . These particles

carry either a positive charge called _____ or a negative charge

called _____ .

— — — — — — — — — — — — — —

ions, cations, anions

3

What is the difference between a cation and an anion? _____

_____ .

– – – – – – – – – – – – – – – –

cation—positive charge and anion—negative charge

Table 1 gives the principal cations and anions in human body fluid. Since we will be referring to these elements and their symbols throughout the program, take a few minutes now to memorize them. Be sure to note the + and − signs. We will not have a separate section on bicarbonate and phosphate in this chapter, but we will be talking about them later, so be sure to remember them. When you think you are ready, go ahead to the frames which follow, referring back to the table only when necessary.

TABLE 1   Cations and Anions

Cations		Anions	
Na^+	(Sodium)	Cl^-	(Chloride)
K^+	(Potassium)	HCO_3^-	(Bicarbonate)
Ca^{++}	(Calcium)	HPO_4^{--}	(Phosphate)
Mg^{++}	(Magnesium)		

4

Place a C in front on the Cations and an A in front of Anions.

____ a. K   ____ e. Na

____ b. Mg   ____ f. Ca

____ c. Cl   ____ g. HPO_4

____ d. HCO_3

– – – – – – – – – – – – – – – –

a. C   e. C
b. C   f. C
c. A   g. A
d. A

5

If you have had Chemistry, these ions and their symbols should be familiar to you. Complete the following chart using proper names and/ or symbols.

Names of Ions	Symbols
Sodium	
	K
Calcium	
	Cl
Bicarbonate	
	HPO_4
	Mg

– – – – – – – – – – – – – – – –

Names	Symbols
Potassium	Na
Chloride	Ca
Phosphate	HCO_3
Magnesium	

6

For electrical balance, the quantities of cations and anions in a solution, expressed in milliequivalents (mEq) always equal each other.

Electrolytes differ in their chemical activity for sodium has one positive charge and calcium has *_____.

– – – – – – – – – – – – – – – –

two positive charges

7

The term milliequivalents is used to express the number of ionic charges

of electrolytes on an equal basis. It measures the _____ activity
of ions or elements. Refer to Chapter 1, Frames 55 and 56 if needed to
explain milliequivalents.

The total cations in milliequivalents must equal the total _____
in milliequivalents.

— — — — — — — — — — — — — — — —

chemical, anions

8

Milliequivalents consider electrolytes in terms of their *_____

_____ rather than their weight.

— — — — — — — — — — — — — — —

chemical activity

9

Electrolytes have different weights, but are considered during therapy
in terms of their activity, which would be expressed as (milliequivalents/
milligrams).

— — — — — — — — — — — — — — —

milliequivalents

Table 2 gives the weights and equivalences of five ions. Note how the
weights of the named ions differ, but the equivalence remain the same
according to their ionic charge. You are not expected to memorize the
weights of these ions.

TABLE 2   Electrolyte Equivalents

Kind of Ion	Weight	Equivalence
Sodium +	23 mg	1 mEq
Potassium +	39 mg	1 mEq
Chloride −	35 mg	1 mEq
Calcium ++	40 mg	2 mEq
Magnesium ++	24 mg	2 mEq

10

An ion with two charges would have the same equivalence as *_____

_____.

— — — — — — — — — — — — — —

another ion with two charges

11

Name a cation and an anion with the same equivalence but different

weights. _____.

— — — — — — — — — — — — — —

Sodium and chloride or potassium and chloride

12

The electrolyte composition of fluid differs within the two main classes
of body fluid.

The two main classes of body water are _____ and

_____ .

— — — — — — — — — — — — — —

intracellular and extracellular

Table 3 gives the ion concentrations in the intravascular fluid (which is
frequently referred to as plasma), interstitial fluid, and intracellular
fluid. Take a few minutes to memorize the fluids and their greatest concentration of ions. Memorizing the numbers is not necessary. Refer back
to the table when necessary.

TABLE 3   Electrolyte Composition of Body Fluid (mEq/L)

| Ions | Extracellular | | Intracellular |
	Intravascular or Plasma	Interstitial	
Na^+	142	145	10
K^+	5	4	141
Ca^{++}	5	3	2
Mg^{++}	2	1	27
Cl^-	104	116	1
HCO_3^-	27	30	10
HPO_4^{--}	2	2	100

13

What are the three principal ions in intravascular fluid? _____,

_____ , and _____ .

What are the three principal ions in interstitial fluid?_____,

_____ , and _____ .

What are the three principal ions in intracellular fluid? _____,

_____ , and _____ .

— — — — — — — — — — — — — — —

sodium or Na, chloride or Cl, and bicarbonate or HCO_3
same as intravascular fluid
potassium or K, phosphate or HPO_4, and magnesium or Mg

Diagram 3 shows the various cations and anions in extracellular and
intracellular fluids. Pay special attention to the principal cations and
anions in these fluids. Refer back to the diagram only when necessary.

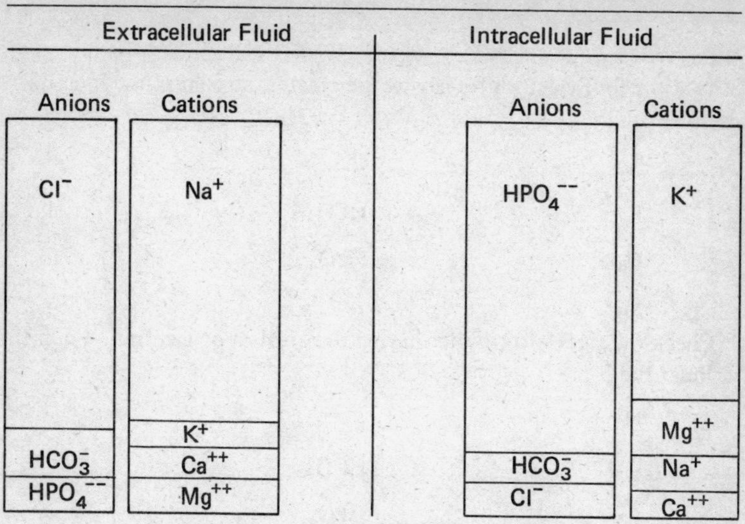

Extracellular Fluid		Intracellular Fluid	

Diagram 3. Anions and cations in body fluid.

14

The principal cation in extracellular fluid is_____ .

The principal cation is intracellular fluid is _____ .

— — — — — — — — — — — — — — —

sodium or Na, potassium or K

15

The principal anion in extracellular fluid is _____ .

The principal anion in intracellular fluid is_____ .

— — — — — — — — — — — — — — —

chloride or Cl, phosphate or HPO_4

16

Check the four electrolytes having the greatest concentration in extra-cellular fluid.

_____ Na   _____ Cl

_____ K   _____ HCO$_3$

_____ Ca   _____ HPO$_4$

_____ Mg

Check the three electrolytes having the greatest concentration in intra-cellular fluid.

_____ Na   _____ Cl

_____ K   _____ HCO$_3$

_____ Ca   _____ HPO$_4$

_____ Mg

- - - - - - - - - - - - - - -

Na   K
Cl   Mg
HCO$_3$ HPO$_4$
Ca

Review

Q. 1

Chemical compounds that emit an electrical charge when placed in so-lutions are known as _____ .

- - - - - - - - - - - - - - -

electrolytes

Q. 2

Dissociated particles of electrolyte which carry either a positive or negative charge are known as_____.

— — — — — — — — — — — — — — —

ions

Q. 3

Define the following terms:
Cation.
Anion.

— — — — — — — — — — — — — — —

positive charged ion, negative charged ion

Q. 4

Place ICF in front of the electrolytes which have their highest concentration in the intracellular fluid and ECF for the electrolytes with their highest concentration in the extracellular fluid.

_____ Na   _____ Cl

_____ K   _____ HCO_3

_____ Ca   _____ HPO_4

_____ Mg

— — — — — — — — — — — — — —

Na = ECF   Cl = ECF
K = ICF HCO_3 = ECF
Ca = ECF HPO_4 = ICF
Mg = ICF

POTASSIUM

17

Although potassium is present in all body fluids, it is found mostly in

_____ fluid.

What kind of ion is potassium? _____ .

_ _ _ _ _ _ _ _ _ _ _ _ _ _ _ _

intracellular, cation

Diagram 4 tells the effect of too much potassium or not enough in our body cells. Memorize the normal range of our serum potassium. You may wonder why the range of serum potassium and not cell potassium, since the cells have the highest concentration of potassium. It is easier to aspirate serum from the intravascular fluid than to aspirate body cells. When you are ready, go ahead to the frames following the diagram and refer back to the diagram when necessary.

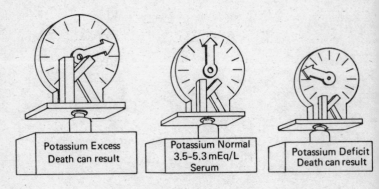

Diagram 4. Potassium—balance and imbalance.

18

The kidneys excrete excess potassium. If the kidneys fail to function, what could result? _____.

– – – – –– – – – – – – – – –

excess of potassium and death

19

Either too much or too little potassium can be _____ .

The normal mEq/L of serum potassium is _____ .

– – – – – – – – – – – – – – – –

fatal, 3.5 – 5.3 mEq/L

20

Potassium is necessary for muscular activity, primarily the cardiac muscle. Potassium strengthens the heart muscle, but having too much or too little potassium will cause cardiac arrest.

How do you think too much potassium can cause cardiac arrest?

_____.

How do you think too little potassium can cause cardiac arrest?

_____.

– – – – – – – – – – – – – – –

It causes irritability of the heart muscle, increasing and then decreasing the heart beat.

Weakening the heart muscle causing it to beat irregularly.

21

Too much serum potassium is known as hyperkalemia and too little serum potassium is known as hypokalemia.

A serum potassium of 3.0 mEq/L would be known as_____.

A serum potassium of 4.2 mEq/L would be known as_____.

A serum potassium of 5.8 mEq/L would be known as _____.

– – – – – – – – – – – – – – – –

hypokalemia, normal, hyperkalemia

22

The assimilative processes involved in the formation of new tissue (the synthesis of complex molecules from simple molecules) are referred to as anabolism and the reactions concerned with tissue breakdown (the breakdown of complex molecules to simple molecules with a release of chemical energy) are referred to as catabolism.

When cellular activity is anabolic (state of building up), potassium will enter the cells. When cellular activity is catabolic (state of breaking down), potassium will leave the cells.

Potassium enters the cells in _____ states and leaves the cells

in _____ states.

– – – – – – – – – – – – – – – –

anabolic, catabolic

23

Potassium may leave the cells under various conditions. When tissues are destroyed as a result of trauma, starvation, or wasting diseases, large

amounts of potassium *_____.

Potassium leaves the cells in _____ states.

– – – – – – – – – – – – – – – –

will leave the cells, catabolic

24

During exercise, when muscles contract, the cells lose potassium and absorb a nearly equal quantity of sodium from the extracellular fluid.

After exercise, when the muscles are recovering from fatigue, potassium reenters the cells and most of the sodium goes back to the extracellular fluid.

During exercise which ion may be more plentiful in the extracellular

fluid—the potassium ion or the sodium ion. _____

_____ .

_____

The potassium ion. Of course it depends on how much exercise. The K ion has to go somewhere so it goes into the extracellular fluid.

25

With exercise, potassium leaves the cells, causing muscular fatigue.

After exercise, potassium *_____ .

Potassium enters the cells in _____ states.

_ _ _ _ _ _ _ _ _ _ _ _ _ _ _

reenters the cells, anabolic

26

The muscles, after releasing potassium from the cells, are like "half-filled water bottles," and are soft.

The soft muscles are a result of _____ .

_ _ _ _ _ _ _ _ _ _ _ _ _ _ _

hypokalemia or potassium loss

27

Name as many conditions as you can under which potassium might leave the cells.

1.

2.

3.

4.

_ _ _ _ _ _ _ _ _ _ _ _ _ _ _

1. trauma
2. exercise
3. starvation
4. wasting disease

28

In stress, which can be caused by a harmful condition or emotional strain, an excessive amount of potassium is lost through the kidneys. The potassium leaves the cells, depleting the cells' needs. From the adrenal gland one of the adrenal cortical hormones, which is aldosterone, is produced in abundance during stress. This hormone will influence the kidneys in excreting potassium.

Frequently the cations, K and Na, have an opposing effect on each other in the extracellular fluid. When one is retained the other is excreted.

Therefore, with an excessive production of aldosterone, what will happen to the cations—K and Na?

_____.

_ _ _ _ _ _ _ _ _ _ _ _ _ _

K will be excreted and Na will be retained

29

Anytime you have an excess potassium in the urine and if the kidney function is normal, then the excessive amount will be excreted.

If kidneys are injured or diseased, then:

(   ) a. The potassium piles up in the extracellular fluid.
(   ) b. The potassium piles up in the intracellular fluid.
(   ) c. The potassium is excreted through the skin.

— — — — — — — — — — — — — — —

a. X
b. __
c. __

Table 4 explains the various clinical conditions that cause too much potassium, or excess, and too little potassium, or deficit. Study this table carefully, noting the potassium changes with various conditions and reasons for these changes. You may have to refer back to this table several times.

TABLE 4 Conditions Causing Potassium Deficit and Excess

Condition	Potassium (K) Deficit— Hypokalemia	Potassium (K) Excess— Hyperkalemia
Diarrhea	K is found in rich quantities in the intestinal secretions. K passes out with the secretions.	
Vomiting	Vomitus contains rich amounts of K.	
Dehydration	Loss of K from cells.	
Diuretics	Large doses of a carbonic anhydrase inhibitor drug will increase excretion of K. Many other diuretics will cause the same loss.	
Burns	K is used in great quantities to repair burn tissues. Burns will cause a reduction of K in the cells.	
Anorexia and starvation	K is found in most food. However, lack of food, or a diet of food lacking in K, will result in a severe K deficit.	
Trauma, injury, surgery	K is used in great quantities to repair injured tissues.	
Stress	Excessive amounts of K are lost through kidneys because of the production of the adrenal cortical hormone, aldosterone.	
Kidney dysfunction		Kidney dysfunction will cause a retention of K in the extracellular fluid.
Adrenal gland disease	Kidney excretion of potassium is accelerated by ACTH and the adrenal cortical hormones, e.g., cortisone and aldosterone.	Lack of adrenal cortical hormone will cause a retention of K and an excretion of Na.

30

What two conditions might cause loss of K from the gastrointestinal

tract? _____.

— — — — — — — — — — — — — —

vomiting and diarrhea

31

Potassium excess or retention will occur with *_____

and *_____ .

— — — — — — — — — — — — — —

kidney dysfunction and adrenal gland insufficiency

32

Trauma and injury to tissues as result of burns and surgery can cause a

potassium (deficit/excess). Why? _____ .

— — — — — — — — — — — — —

deficit, Great quantities of K are needed to repair tissues

33

Dehydration, diuretics, and starvation will result in a *_____

_____ .

— — — — — — — — — — — — — —

potassium deficit

34
List the clinical conditions causing a K deficit.
1.
2.
3.
4.
5.
6.

_ _ _ _ _ _ _ _ _ _ _ _ _ _ _

1. diarrhea, vomiting
2. dehydration, starvation
3. diuretics
4. burns
5. trauma or injury
6. surgery
7. stress
8. increase of adrenal cortical hormones

35
Though 98% of potassium is found in cells, focus is placed on the extra-cellular fluid for it is more readily available to study.

The normal mEq/L of serum potassium (in extracellular fluid) is

_____ .

_ _ _ _ _ _ _ _ _ _ _ _ _ _

3.5 – 5.3 mEq/L

Table 5 gives the signs and symptoms associated with hyper-hypo-kalemia. You should memorize the symptoms most commonly seen, which are marked with double asterisks. You should also become familiar with the other symptoms for these could be related to patients when the commonly seen symptoms occur. This would aid the physician in making a more positive diagnosis. Please learn this table and refer back to it as needed. Patients with hyperkalemia and hypokalemia can be found in many of your clinical situations. You may save a patient's life by recognizing and associating their symptoms as one of the potassium imbalances. If you are not familiar with these words, refer to the glossary.

TABLE 5   Signs and Symptoms Related to Hyper-Hypokalemia

Classes	Hyperkalemia	Hypokalemia
Gastrointestinal abnormalities	*Nausea *Diarrhea **Abdominal cramps	*Anorexia Nausea *Vomiting Diarrhea **Abdominal distention
Cardiac abnormalities	**Tachycardia, later bradycardia Cardiac arrest	**Arrhythmia **Dizziness Cardiac arrest
Urinary abnormalities	**Oliguria or anuria	Polyuria
Neurological abnormalities	Weakness, numbness or tingling	**Malaise **Muscular weakness Mental depression Drowsiness Confusion Paralysis of respiration
Amount in extra-cellular fluid	Above 5.3 mEq/L	Below 3.5 mEq/L

**Most commonly seen symptoms of hyper-hypokalemia.
*Commonly seen symptoms of hyper-hypokalemia.

36

With hyperkalemia, the heart beats very fast, which is known as tachy-cardia, and then it slows down, known as bradycardia. The heart goes into a block with little or no impulses transmitted and then cardiac arrest, which results in death.

You recall that the kidneys are responsible for excreting excessive amounts of potassium not needed by the body. If the kidneys excrete a small amount of urine, known as oliguria, or no urine, known as anuria,

what could occur to the serum potassium level? _____ .

_ _ _ _ _ _ _ _ _ _ _ _ _ _ _ _ _

increase, or rise, or hyperkalemia

37

Name the three most commonly seen symptoms of hyperkalemia.

_____ , _____ , and _____ .

_ _ _ _ _ _ _ _ _ _ _ _ _ _ _ _ _

Abdominal cramps, tachycardia and later bradycardia, and oliguria or anuria

38

Can you name the other symptoms related to hyperkalemia?
1.
2.
3.
4.

_ _ _ _ _ _ _ _ _ _ _ _ _ _ _ _

1. nausea
2. diarrhea
3. weakness
4. numbness or tingling

39

It may be dangerous if the amount of potassium in extracellular fluid is

above _____ .

— — — — — — — — — — — — — — —

5.3 mEq/L

40

Hypokalemia causes the muscle to be soft, like half-filled water bottles, and weak. The abdomen becomes bloated due to smooth muscle weakness and not due to flatus. The blood pressure goes down (hypotension) and dizziness occurs. Malaise or uneasiness occurs.

The heat beat is irregular known as _____ . Eventually if the irregularity of the heart beat is not corrected, bradycardia occurs and then cardiac arrest.

— — — — — — — — — — — — — — —

arrhythmia

41

Polyuria or excess urine output can be a symptom of hypokalemia. State two causes of hypokalemia in which polyuria is involved. (Refer to

Table 4 if needed.) _____ and _____ .

— — — — — — — — — — — — — — —

excess aldosterone and diuretics

42

Name the five most commonly seen symptoms of hypokalemia.

_____ , _____ , _____ ,

_____ , and _____ .

— — — — — — — — — — — — — —

abdominal distention, dizziness, arrhythmia, malaise, muscular weakness

43

Can you name the other symptoms related to hypokalemia?
1.
2.
3.
4.
5.
6.
7.

— — — — — — — — — — — — —

1. anorexia
2. nausea
3. vomiting
4. diarrhea
5. polyuria
6. mental depression
7. drowsiness
8. confusion
9. paralysis of respiration

44

It may be dangerous if the amount of potassium in the extracellular

fluid is below _____ .

— — — — — — — — — — — — —

3.5 mEq/L

45

A weak grip, an irregular pulse, and dizziness upon standing could be

signs of _____ .

— — — — — — — — — — — — —

hypokalemia

Clinical Considerations

46

Digitalis is a drug that strengthens the heart muscle and slows down the heart beat. It is believed that digitalis excretes potassium from the heart muscle which is needed for heart contraction. What type of

potassium imbalance would occur? _____.

Name the most common symptoms of this imbalance.
1.
2.
3.
4.
5.

– – – – – – – – – – – – – – –

hypokalemia
1. abdominal distention
2. dizziness
3. arrhythmia
4. malaise
5. muscular weakness

47

Bizarre cardiac arrhythmia in digitalized patients can occur due to

hypokalemia. Arrhythmia is a major symptom of _____.

– – – – – – – – – – – – – –

hypokalemia

48

An ordinary dose of digitalis with a low serum potassium may induce symptoms of digitalis toxicity since this deficit makes the heart sensitive to digitalis. Symptoms of digitalis toxicity consist of bradycardia, which is a slow heart beat, and/or arrhythmia.

This toxic reaction is due to *_____ .

However, if an individual is taking large doses of digitalis over a prolonged period of time, digitalis toxicity can occur due to hypokalemia.

Can you explain why the hypokalemia in this situation? _____

_____ .

Give two causes of digitalis toxicity:
1.
2.

_ _ _ _ _ _ _ _ _ _ _ _ _ _ _

a drop in body potassium
Excessive amounts of digitalis will excrete potassium.
1. A patient receiving ordinary doses of digitalis and having low serum K.
2. A patient receiving large doses of digitalis causing hypokalemia.

49

In the cirrhotic patient, having degenerated liver cells, hypokalemia can precipitate hepatic coma or liver failure.

As a nurse caring for a patient with cirrhosis, you would alert the physician of any low serum K levels, and you would watch for symptoms of *_____ .

_ _ _ _ _ _ _ _ _ _ _ _ _ _ _

hypokalemia—also hepatic coma

50

Eighty to ninety percent of potassium excretion is lost in the urine and only a very small percentage is lost in the feces.
Which of the following would have the greater loss of potassium?
( ) a. An individual taking a laxative.
( ) b. An individual taking a diuretic.

– – – – – – – – – – – – – – –

a. –
b. X

51

Hyperglycemia, an increased blood sugar, is a symptom of diabetes mellitus. It will cause a failure of cells to utilize glucose, permitting potassium

to leave the cells. Do you know why? _____.
   When cells do not receive their proper nutrition, catabolism occurs and potassium leaves the cells. Therefore, in severe hyperglycemia (hypokalemia/hyperkalemia) will occur.

– – – – – – – – – – – – – – –

There is not enough insulin to utilize glucose, hypokalemia

52

Rapid correction of abnormal cellular metabolism in a diabetic patient by administering glucose and insulin may lead to rapid transfer of potassium from the extracellular fluid to the cell. Therefore, the serum potassium would rapidly (increase/decrease).

– – – – – – – – – – – – – –

decrease

53

When oliguria develops because of poor renal function, potassium is no longer excreted, which results in a high level of serum potassium.

If there is poor renal function, do you think potassium should be administered? _____ .

Why? _____ .

— — — — — — — — — — — — — — —

No—NEVER with poor renal function. Why? Hyperkalemia could be brought to a dangerous level.

54

In cases of kidney failure or renal shutdown, hyperkalemia is frequently the immediate cause of death.

Can you give one reason why kidney failure or renal shutdown might be a cause of death? _____ .

— — — — — — — — — — — — — — —

accumulation of K causing severe hyperkalemia

55

Potassium therapy should not be administered to patients with un-treated adrenal insufficiency and * _____ .

— — — — — — — — — — — — — — —

renal failure or poor renal function

56

Severe serum hyperkalemia may occur from administering intravenous potassium in solution rapidly, not giving time for potassium to pass into the cells.

The normal rate of intravenous flow for potassium is 40 mEq in a quart or liter in 8 hours.

What might result from administering 40 mEq of potassium every 2

hours?_____. Why? _____

_____.

– – – – – – – – – – – – – –

hyperkalemia—not enough time for K to pass into cells

57

With prolonged hypokalemia, circulatory failure and myocardial necrosis, which is heart muscle destruction, can occur. The electrocardiogram would frequently show a flat or inverted T wave. Potassium strengthens the heart muscle, but too little potassium will cause

*_____.

– – – – – – – – – – – – – –

cardiac failure

Diagrams 5 and 6 note electrocardiographic changes found with hypo-hyperkalemia. Students who have had a physiology course and/or a basic knowledge of electrocardiogram, also known as ECG or EKG, will find these diagrams most useful in their own clinical application when monitoring patients. Students who do not have this basic knowledge should refer to a physiology text and/or a text on electrocardiography. You may need this understanding of electrocardiographic changes in the future. Students who do not need this information to practice nursing may skip to the Review following Frame 65.

A brief resumé of the electrocardiogram. The ECG measures the electrical activity from various areas of the heart and records this as P, QRS, and T waves.

P wave measures the electrical activity initiating contraction of the atrium or the atrial muscle.

QRS wave complex measures the electrical activity initiating contraction of the ventricle, which is the thickest part of the heart muscle responsible for forcing blood from the heart into the circulation. "Heart attack," also known as myocardial infarction, frequently affects this part of the heart muscle.

T wave is the electrical recovery of the ventricles.

Abnormal potassium levels affect the T wave of the electrocardiogram. Note the normal T wave structure in Diagram 5 and compare the normal with the abnormal, with Patterns 1 and 2. Study this diagram and then proceed to the frames.

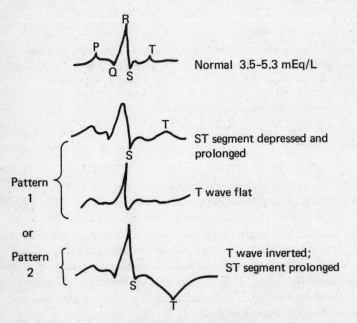

Normal 3.5–5.3 mEq/L

ST segment depressed and prolonged

T wave flat

Pattern 1

or

Pattern 2

T wave inverted; ST segment prolonged

Diagram 5. Electrocardiographic changes in serum potassium deficit. (Adapted from Harry Statland: *Fluid and Electrolytes in Practice,* J. B. Lippincott Co., Philadelphia, Pa., 1963, p. 120.)

58

Name the abnormal changes of the T wave comparing with the normal.

_____ and _____ .

The abnormal T waves would indicate a potassium (excess/deficit).

_ _ _ _ _ _ _ _ _ _ _ _ _ _ _

flat T wave and inverted T wave, deficit

59

The ST segment is prolonged in both the patterns. This change also relates to * _____ .

_ _ _ _ _ _ _ _ _ _ _ _ _ _ _

potassium deficit

60

With serum potassium deficit the following electrocardiographic changes could occur:

(   ) a. Flat T wave
(   ) b. Inverted T wave
(   ) c. High-peaked T wave
(   ) d. ST segment depressed and prolonged
(   ) e. Absence of the P wave

_ _ _ _ _ _ _ _ _ _ _ _ _ _ _

a. X
b. X
c. —
d. X
e. —

High-peaked T waves are an early electrocardiographic sign of hyperkalemia. Heart block can result from severe hyperkalemia, e.g., 8 to 10 mEq/L of serum potassium. Study Diagram 6 carefully noting especially the T waves, QRS complex and P wave. If any of the words are unfamiliar please refer to a physiology text and/or a text on electrocardiography.

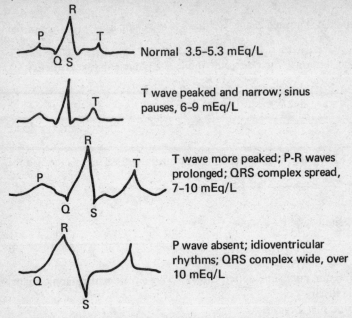

P R Q S T — Normal 3.5–5.3 mEq/L

T — T wave peaked and narrow; sinus pauses, 6–9 mEq/L

P R Q S T — T wave more peaked; P-R waves prolonged; QRS complex spread, 7–10 mEq/L

R Q S — P wave absent; idioventricular rhythms; QRS complex wide, over 10 mEq/L

Diagram 6. Electrocardiographic changes in serum potassium concentration. Changes that do occur are most marked in the precordial leads over the right side (V_1–V_4 position) of the heart. (Adapted from Harry Statland: *Fluid and Electrolytes in Practice*, J. B. Lippincott Co., Philadelphia, Pa., 1963, p. 116.)

61
Name the abnormal change of the T wave occurring with hyperkalemia.

_____ .

– – – – – – – – – – – – – – – –

high-peaked T wave

ANSWER SHIELD

FLUIDS AND ELECTROLYTES
WITH
CLINICAL APPLICATIONS

A Programmed Approach

Joyce LeFever Kee

Assistant Professor, Medical-Surgical Nursing
College of Nursing, University of Delaware
Newark, Delaware

John Wiley & Sons, Inc.   New York / London / Sydney / Toronto

62

A flat or inverted T wave on an electrocardiogram frequently indicates

a _____ state, whereas a high-peaked T wave could

indicate a _____ state.

– – – – – – – – – – – – – – – –

hypokalemic, hyperkalemic

63

As the serum potassium concentration increases, 7–10 mEq/L, the T

wave becomes_____ , the P-R waves are_____ ,

and the QRS complex is _____ .

– – – – – – – – – – – – – – – –

more peaked, prolonged, spread

64

The following electrocardiographic changes could occur with a high
serum potassium:

( ) a. Flat T wave
( ) b. Inverted T wave
( ) c. High-peaked T wave
( ) d. ST segment depressed and prolonged
( ) e. QRS complex spread
( ) f. P-R waves prolonged

– – – – – – – – – – – – – – – –

a. –
b. –
c. X
d. –
e. X
f. X

65

Place KC for potassium concentration or KD for potassium deficit which would cause the following electrocardiographic changes.

_____ a. Flat T wave

_____ b. Inverted T wave

_____ c. High-peaked T wave

_____ d. ST segment depressed and prolonged

_____ e. QRS complex spread

_____ f. P-R waves prolonged

_ _ _ _ _ _ _ _ _ _ _ _ _ _ _

a. KD
b. KD
c. KC
d. KD
e. KC
f. KC

Review

Q. 1

Potassium is a(n) (cation/anion) found primarily in the (extracellular/intracellular) fluid.

_ _ _ _ _ _ _ _ _ _ _ _ _ _ _

cation, intracellular

Q. 2

A high serum potassium concentration is known as _____,

whereas a low serum potassium concentration is known as

_____ .

_ _ _ _ _ _ _ _ _ _ _ _ _ _ _

hyperkalemia, hypokalemia

Q. 3

What can be the result of too much or too little potassium? _____ .

_ _ _ _ _ _ _ _ _ _ _ _ _ _ _ _

death

Q. 4

Which of the following clinical conditions could cause potassium to leave the cells?

( )   a. Trauma
( )   b. Exercise
( )   c. Starvation
( )   d. Stress
( )   e. Sleeping
( )   f. Thinking
( )   g. Burns
( )   h. Overproduction of adrenal cortical hormones
( )   i. Adrenal insufficiency
( )   j. Wasting diseases
( )   k. Poor renal function

_ _ _ _ _ _ _ _ _ _ _ _ _ _ _ _

a. Trauma
b. Exercise
c. Starvation
d. Stress
g. Burns
h. Overproduction of adrenal cortical
j. Wasting diseases

Q. 5

Vomiting and diarrhea may cause a potassium (excess/deficit).

_ _ _ _ _ _ _ _ _ _ _ _ _ _ _ _

deficit

Q. 6
Poor renal function and adrenal insufficiency will cause a potassium (excretion/retention).

_ _ _ _ _ _ _ _ _ _ _ _ _ _

retention

Q. 7
The considered "normal" serum potassium is _____ mEq/L.

_ _ _ _ _ _ _ _ _ _ _ _ _ _

3.5 – 5.3

Q. 8
List at least 3 common symptoms found with hyperkalemia.
1.
2.
3.

_ _ _ _ _ _ _ _ _ _ _ _ _

Abdominal cramps
Tachycardia
Nausea, diarrhea

Q. 9
List at least 3 common symptoms found with hypokalemia.
1.
2.
3.

_ _ _ _ _ _ _ _ _ _ _ _ _ _

Abdominal distention
Muscular weakness
Arrhythmia
Dizziness
Anorexia
Diarrhea

Q. 10

The heart needs potassium for (relaxation/contraction).

— — — — — — — — — — — — — —

contraction

Q. 11

High-peaked T waves are frequently an early electrocardiographic sign of (hypokalemia/hyperkalemia). Flat or inverted T waves are a sign of (hypokalemia/hyperkalemia).

— — — — — — — — — — — — — —

hyperkalemia, hypokalemia

Q. 12

Hyperglycemia in individuals with diabetes mellitus results in a failure of

*_____causing

potassium to *_____ .

— — — — — — — — — — — — — —

cells utilizing glucose, leave the cells

Q. 13

Potassium leaves the cells in a(an) _____ state.

— — — — — — — — — — — — — —

catabolic

Q. 14

Rapidly administered intravenous potassium can cause a severe

_____ . Why? _____.

— — — — — — — — — — — — — —

hyperkalemia
Not enough time is given for potassium to pass into the cells.

SODIUM

66
Sodium is the main cation found in the_____
fluid.

_ _ _ _ _ _ _ _ _ _ _ _ _ _ _ _

extracellular or intravascular

67
Sodium loss from the skin is negligible under normal conditions, but
with increased environmental temperature, fever, and/or muscular exercise, the loss rises.
    If an individual runs a race and the atmospheric temperature is 100,
what do you think would happen to the sodium in his body?

_____.

_ _ _ _ _ _ _ _ _ _ _ _ _ _ _ _

sodium loss

68
The normal concentration of sodium in the extracellular fluid is 135 –
146 mEq/L.
    The normal concentration of sodium in perspiration is 50 – 100 mEq/L,

which is about half the concentration found in the *_____.

_ _ _ _ _ _ _ _ _ _ _ _ _ _ _ _

extracellular fluid

69
When the body's sodium level is elevated, perspiration is not a means of
regulating sodium excretion, for it is regarded as a by-product of temperature regulation.
    The concentration of sodium in the extracellular fluid is _____.

_ _ _ _ _ _ _ _ _ _ _ _ _ _ _ _

135 – 146 mEq/L

70

The concentration of sodium in perspiration is_____.
   Explain why perspiration is not a means of regulating sodium excretion. _____.

– – – – – – – – – – – – – –

50 –100 mEq/L, It is a by-product of temperature regulation.

71

Bones contain as much as 800 to 1000 mEq/L of sodium, but only a portion of the sodium is available for exchange with sodium in other parts of the body.
   The concentration of sodium in the extracellular fluid is_____.

– – – – – – – – – – – – – –

135 – 146 mEq/L

72

Bones contain (more/less) sodium than extracellular fluid.

– – – – – – – – – – – – – –

more

73

Thirst often leads to the replacement of water, but not sodium.
   One (can/cannot) replace sodium by drinking lots of water.

– – – – – – – – – – – – – –

cannot

74

Ocean water is about three times as salty as our body fluid—far too salty for our body organs, i.e., stomach and intestine.
   Ocean water would be considered a (hypotonic/hypertonic) fluid.

– – – – – – – – – – – – – –

hypertonic

75

Therefore, in cases of ocean water ingestion, the water is drawn from the body fluid into the stomach and intestines by the process of (osmosis/ diffusion).

— — — — — — — — — — — — — —

osmosis

76

As the stomach and intestines accumulate huge volumes of water, vomiting then occurs. Explain the effect of vomiting due to ocean water ingestion in relation to the body water and serum sodium.

_____.

— — — — — — — — — — — — —

body water deficit and high serum sodium

77

An elevated serum sodium is known as sodium excess or hypernatremia and a decreased serum sodium is known as sodium deficit or hyponatremia.

Hypernatremia is known as _____. Hyponatremia is

known as _____.

— — — — — — — — — — — — — —

sodium excess, sodium deficit

78

One of the main functions of sodium is to influence the distribution of water in the body. Water will accompany sodium.

A name for sodium excess is _____.

A name for sodium deficit is _____.

A function of sodium is to influence the distribution of *_____

_____ .

_ _ _ _ _ _ _ _ _ _ _ _ _ _ _ _

hypernatremia, hyponatremia, water in the body. Water accompanies sodium.

Table 6 explains how one organ and two glands influence serum sodium. The kidneys have an important role in maintaining homeostasis of body sodium. The posterior hypophysis or posterior pituitary gland secretes ADH which will absorb large quantities of water and sodium from the kidneys. The adrenal glands are composed of two sections, the cortex and the medulla, each secreting their own hormones. The hormones from the adrenal cortex are frequently referred to as steroids. Study this table carefully. Refer to a physiology text for any further clarification.

TABLE 6   Influences Affecting Serum Sodium

Organ	
Kidneys	The regulators. Kidneys maintain homeostasis through excretion or absorption of sodium from the renal tubules according to the excess or deficit of serum sodium.
Glands	
1. Posterior hypophysis or posterior pituitary gland	Pituitary antidiuretic hormone (ADH) favors water absorption from the distal tubules of the kidneys and thus sodium excretion is promoted to a lesser extent.
2. Adrenal cortex of the adrenal glands	The adrenal cortical hormones, e.g., cortisone and aldosterone, favor sodium absorption from the renal tubules. These steroids influence the kidneys to absorb sodium and excrete potassium.

79
The chief regulation of sodium occurs within the_____.

- - - - - - - - - - - - - - - -

kidneys

80
The amount of water absorbed from the kidneys depends upon the amount of ADH being secreted. If less water is absorbed from the renal

tubules, what happens to the sodium?_____.

- - - - - - - - - - - - - - -

less sodium excretion

81
Explain the effect of cortisone and aldosterone on sodium and potas-

sium._____.

- - - - - - - - - - - - - - -

They influence the kidneys to absorb sodium and excrete potassium.

82
Renal impairment and circulatory impairment result in reduced glomer-ular filtration (that is, filtration through the renal capillaries known as glomeruli) which causes sodium retention.
    Normal kidney function is necessary to conserve or excrete water and

_____.

- - - - - - - - - - - - - - -

sodium

83
Impairment of renal and circulatory function will cause sodium (reten-tion/excretion).

- - - - - - - - - - - - - - -

retention

84

Large amounts of sodium are contained within the following body secretions—saliva, gastric secretions, bile, pancreatic juice, and intestinal secretions.

Indicate which of the following body secretions contain large quantities of sodium:

( ) a. Saliva
( ) b. Thyroid secretions
( ) c. Gastric secretions
( ) d. Bile
( ) e. Parathyroid secretions
( ) f. Pancreatic juice
( ) g. Intestinal secretions

_ _ _ _ _ _ _ _ _ _ _ _ _ _ _

a. X
b. –
c. X
d. X
e. –
f. X
g. X

85

Sodium is rapidly shifted back and forth between the intravascular and interstitial spaces.

The shift in sodium from the extracellular to the intracellular space is very slow. When would this shift occur?_____.

_ _ _ _ _ _ _ _ _ _ _ _ _ _ _

when potassium leaves the cells

Table 7 lists specific conditions causing either a sodium deficit or excess or both deficit and excess. If you are not familiar with any words in this table, refer to a physiology text or a nursing text or the glossary. Study this table carefully. Refer to this table as needed.

TABLE 7 Conditions Causing Sodium Deficit and Sodium Excess

Conditions	Na Deficit (Hyponatremia)	Na Excess (Hypernatremia)
Vomiting	Sodium concentration is high in the gastric mucus. Vomiting will decrease the sodium concentration in the alimentary tract.	With severe vomiting, the loss of water is greater than the loss of Na. The saltiness of the remaining body fluids becomes dangerously high.
Sweating, increased environmental temperature, fever, muscular exercise	Large amounts of sodium are lost from the skin, which frequently results from increased environmental temperature, fever, muscular exercise, and sweating.	
Diarrhea	There is a loss of Na from the intestinal secretions.	When babies have diarrhea, their loss of water is greater than their loss of Na.
Tap water enema	The tap water enemas wash out salty intestinal contents from the body.	
Burns	Great quantities of sodium are lost from the burn wounds and from the oozing of the burn surface.	
Surgery	There is loss of Na from postoperative wound drainage, bleeding, and vomiting.	
Gastric suction	A tube inserted through the nose or mouth into the stomach and intestines for purpose of drainage will cause the "salty" gastric and intestinal secretions to pass out the gastrointestinal tract.	
Water intake (excess or decrease)	Drinking great quantities of plain water will dilute extracellular fluid. This is especially significant in burns, surgery, heat prostration in which Na is lost and the remaining body sodium is diluted.	Even when there is a slight Na deficit and no water intake, the Na in the extracellular fluid is greater than normal.

TABLE 7 (continued)

Conditions	Na Deficit (Hyponatremia)	Na Excess (Hypernatremia)
Adrenal cortex dysfunction	Adrenal cortical hormones, e.g., cortisone, and aldosterone, have a sodium-retaining effect on the renal tubules. Adrenal insufficiency will cause a loss of Na across the renal tubules. Addison's Disease is an example of adrenal insufficiency.	Excessive adrenal cortical hormone will cause an excess of sodium in the body. Cushing's Syndrome is an example of an overproduction of adrenal cortical hormone.
Kidney, heart, and circulatory dysfunctions	With advanced renal disorders, the renal tubules do not respond to the antidiuretic hormone; therefore, there is a loss of sodium and water.	Reduced glomerular filtration will result in Na retention. Congestive heart failure, shock, or obstruction of the arterial supply to the kidney will result in Na retention.

86

Place a D for sodium deficit or E for sodium excess concerning the following:

( ) a. Vomiting
( ) b. Tap water enema
( ) c. Increased environmental temperature fever and muscular exercise
( ) d. Decrease of adrenal cortical hormone
( ) e. Increase of adrenal cortical hormone
( ) f. Burns
( ) g. Great volumes of water intake
( ) h. Little to no water intake
( ) i. Sweating
( ) j. Surgery
( ) k. Gastric suction
( ) l. Diarrhea
( ) m. Severe diarrhea in babies
( ) n. Reduced glomerular filtration (kidney dysfunction)
( ) o. Congestive heart failure and shock

— — — — — — — — — — — — — — — ⌐

a. D	i. D
b. D	j. D
c. D	k. D
d. D	l. D
e. E	m. E
f. D	n. E
g. D	o. E
h. E	

87

Vomiting will (increase/decrease) the sodium concentration in the alimentary tract.

— — — — — — — — — — — — — — —

decrease

88

Name the conditions causing a large sodium loss through the skin.
1.
2.
3.
4.

— — — — — — — — — — — — — — —

1. sweating
2. increased environmental temperature
3. fever
4. muscular exercise

89

What effects do cortisone and aldosterone have on the renal tubules?

_____.

— — — — — — — — — — — — — —

sodium retention and potassium excretion

90

Addison's Disease occurs when there is an adrenal cortical hormone
(insufficiency/overproduction), whereas Cushing's Syndrome occurs
when there is an adrenal cortical hormone (insufficiency/overproduc-
tion).

— — — — — — — — — — — — — —

insufficiency, overproduction

91

With Addison's Disease, there is a sodium _____.

With Cushing's Syndrome, there is a sodium _____.

— — — — — — — — — — — — — —

loss, retention

92

With burns, there is a great sodium _____. Why? _____

_____.

_ _ _ _ _ _ _ _ _ _ _ _ _ _

loss, due to oozing at the burn surface (or because sodium replaces potassium in the damaged cells).

93

Repeated tap water enemas can result in a * _____. Why?

_____.

_ _ _ _ _ _ _ _ _ _ _ _ _ _

sodium loss, They wash away salty intestinal secretions.

94

Postoperative wound drainage, bleeding, and vomiting can cause a sodium (deficit/retention).

_ _ _ _ _ _ _ _ _ _ _ _ _ _

deficit

95

The use of gastric suction for the purpose of drainage can cause

(hypernatremia/hyponatremia). Why? _____.

_ _ _ _ _ _ _ _ _ _ _ _ _ _

hyponatremia, Gastric and intestinal secretions pass out through the tube.

96

Excessive intake of plain water as a result of thirst due to burns, surgery, and heat prostration could cause a (hypernatremic/hyponatremic) state.

Why? _____.

_ _ _ _ _ _ _ _ _ _ _ _ _ _

hyponatremic, It will dilute extracellular fluid.

97

Kidney, heart, and circulatory dysfunction can cause occurrence of (hypernatremia/hyponatremia).

Why? _____.

_ _ _ _ _ _ _ _ _ _ _ _ _ _ _ _

hypernatremia, Renal tubules do not respond to ADH.

98

Congestive heart failure, shock, or obstruction of the arterial supply to

the kidney will cause sodium _____ . Why?_____

_____ .

_ _ _ _ _ _ _ _ _ _ _ _ _ _ _ _

retention, Reduced glomerular filtration

Table 8 gives the signs and symptoms associated with hypo-hyper-natremia. You should memorize the symptoms commonly seen, which are marked with an asterisk. Specific gravity may be a new word to you. It is a weight of a substance, e.g., urine, in comparison with an equal volume of water. Water has a specific gravity of 1.000. Study this table carefully. Refer to the glossary for any unknown words and refer back to this table as needed.

TABLE 8   **Signs and Symptoms Related to Hypo-Hypernatremia**

Class	Hyponatremia	Hypernatremia
Central nervous system or CNS, muscular and skin	*Abdominal cramps *Muscular weakness Apprehension *Headache Convulsion	*Flushed skin *Elevated body temperature Excitement
Gastrointestinal or GI	*Nausea *Vomiting	*Rough, dry tongue
Cardiac		Blood pressure decrease *Tachycardia
Specific gravity of urine	1.010 ↓ decrease	1.025 ↑ increase
mEq/L	135 and lower	146 and higher

*Commonly seen symptoms of hypo-hypernatremia.

99

Name the commonly seen symptoms of hyponatremia.
1.
2.
3.
4.
5.

— — — — — — — — — — — — — —

1. abdominal cramps
2. muscular weakness
3. headache
4. nausea
5. vomiting

100

It may be dangerous if the amount of sodium in the extracellular fluid

is below _____ .

– – – – – – – – – – – – – – –

135 mEq/L

101

What would the specific gravity of urine be to indicate hyponatremia?

_____ .

– – – – – – – – – – – – – – –

below 1.010

102

Name the commonly seen symptoms of hypernatremia.
1.
2.
3.
4.
Later in this program when you are studying dehydration, you will
notice many symptoms of hypernatremia resemble the symptoms of
dehydration.

– – – – – – – – – – – – – – –

1. flushed skin
2. elevated body temperature
3. rough, dry tongue
4. tachycardia

103

It may be dangerous if the amount of sodium in the extracellular fluid

is above _____ .

– – – – – – – – – – – – – – –

146 mEq/L

104
What would the specific gravity of urine be to indicate hypernatremia?

_____.

– – – – – – – – – – – – – – –

above 1.025

105
Place D for symptoms of sodium deficit and E for symptoms of sodium excess:

( ) a. Abdominal cramps
( ) b. Muscular weakness
( ) c. Flushed skin
( ) d. Headache
( ) e. Elevated body temperature
( ) f. Rough, dry tongue
( ) g. Tachycardia
( ) h. Nausea and vomiting
( ) i. Serum Na, 135 mEq/L and below
( ) j. Serum Na, 146 mEq/L and above

– – – – – – – – – – – – – – –

a. D   f. E
b. D   g. E
c. E   h. D
d. D   i. D
e. E   j. E

Clinical Considerations

106

The majority of laboratory analyses of electrolyte content are carried out on the plasma which represents less than one-twelfth of the total body fluid, therefore, results may occasionally be misleading.

The reason for the use of plasma instead of other body fluid and cells

is that *_____.

– – – – – – – – – – – – – – – –

it is easier to obtain, or a similar response.

107

Explain why electrolyte determinations of plasma may misrepresent

body conditions or illnesses._____.

– – – – – – – – – – – – – – –

Plasma represents less than one-twelfth of the total body fluid.

108

You may have a cardiac patient who has edema and yet his serum sodium concentration is reduced. How do you think this could occur?

_____

_____.

– – – – – – – – – – – – – – –

He would be on a low sodium diet and his water intake would initially dilute Na concentration.

109

If your patient is vomiting following a surgical intervention and is receiving dextrose and water intravenously, one may expect a sodium (excess/deficit) if vomiting persists.

With a patient having severe vomiting without water replacement, one

may expect a sodium (excess/deficit). Why? _____

_____ .

_ _ _ _ _ _ _ _ _ _ _ _ _ _ _ _

deficit, excess, The loss of water would be greater than the sodium

110

If a feeble or debilitated patient receives numerous tap water enemas for the purpose of cleaning the bowel, the enemas could cause a sodium

_____ .

_ _ _ _ _ _ _ _ _ _ _ _ _ _ _

loss

111

Diarrhea can cause either sodium deficit or sodium excess. Babies having

diarrhea could lose more _____ than the ion, sodium; therefore, a

sodium _____ would result.

_ _ _ _ _ _ _ _ _ _ _ _ _ _ _

water, excess

112

Patients receiving steroids, such as cortisone or aldosterone, should be

cautioned in using excess salt. Why _____ .

_ _ _ _ _ _ _ _ _ _ _ _ _ _

steroids have a sodium-retaining effect.

113

With congestive heart failure, there would be a sodium (retention/ excretion). Why _____

_____.

— — — — — — — — — — — — — —

retention, Poor circulation reduces glomerular filtration, therefore, sodium is retained.

Review—Sodium

Q. 1

Sodium is the main (anion/cation) found in the extracellular fluid.

— — — — — — — — — — — — — — —

cation

Q. 2

When does sodium enter the cell? (It is frequently referred to as the sodium and potassium exchange.)

— — — — — — — — — — — — — —

During cell catabolism, K leaves the cell and Na enters.

Q. 3

Name at least three body secretions rich in sodium.
1.
2.
3.

— — — — — — — — — — — — — —

1. saliva
2. bile
3. gastric secretions
4. intestinal secretions
5. pancreatic juice.

Q. 4
The normal concentration of sodium in the extracellular fluid is

_____ .

— — — — — — — — — — — — — — —

135 – 146 mEq/L

Q. 5
The chief regulation of sodium occurs within the_____.

— — — — — — — — — — — — — —

kidneys

Q. 6
The adrenal cortical hormones favor sodium (reabsorption/excretion) and potassium (reabsorption/excretion).

— — — — — — — — — — — — — — —

reabsorption, excretion

Q. 7
Circulatory impairment results in reduced glomerular filtration which causes sodium (retention/excretion).

— — — — — — — — — — — — — —

retention

Q. 8

Place a D where there is a sodium deficit and an E where there is a
sodium excess:

( ) a. Vomiting
( ) b. Tap water enema
( ) c. Increased environmental temperature, fever, and muscular
      exercise
( ) d. Decrease of adrenal cortical hormone
( ) e. Increase of adrenal cortical hormone
( ) f. Burns
( ) g. Great volumes of water intake
( ) h. Little to no water intake
( ) i. Sweating
( ) j. Surgery
( ) k. Gastric suction
( ) l. Diarrhea
( ) m. Severe diarrhea in babies
( ) n. Reduced glomerular filtration (kidney dysfunction)
( ) o. Congestive heart failure and shock

— — — — — — — — — — — — — — — —

a. D	i. D
b. D	j. D
c. D	k. D
d. D	l. D
e. E	m. E
f. D	n. E
g. D	o. E
h. E	

Q. 9

Why might electrolyte determinations of plasma misrepresent conditions

in other spaces?_____.

— — — — — — — — — — — — — — —

It represents less than one-twelfth of the total body fluid.

Q. 10

Place D for symptoms of sodium deficit and E for symptoms of sodium excess:

( ) a. Nausea and vomiting
( ) b. Tachycardia
( ) c. Serum Na, 121 mEq/L
( ) d. Serum Na, 154 mEq/L
( ) e. Rough, dry tongue
( ) f. Elevated body temperature
( ) g. Abdominal cramps
( ) h. Muscular weakness
( ) i. Flushed skin
( ) j. Headache

- - - - - - - - - - - - - - -

a. D   f. E
b. E   g. D
c. D   h. D
d. E   i. E
e. E   j. D

CALCIUM

114

Refer to Table 3 or Diagram 3 if needed. Calcium is a(n) (anion/cation) found in intracellular fluid. Which fluid has the greatest calcium con-

centration?_____.

- - - - - - - - - - - - - - -

cation, extracellular

115

Calcium is a durable chemical substance of the body which is the last element to find its place in the adult body composition and the last element to leave after death.

The element for the preservation of bony remains of dead creatures

and also responsible for the X-ray photograph of bones is_____.

– – – – – – – – – – – – – – – –

calcium

116

Refer to Table 3. Calcium concentration in the blood serum (plasma) is

_____ whereas in the cells it is_____.

– – – – – – – – – – – – – – –

5 mEq/L, 2 mEq/L or a trace

117

Calcium in the extracellular fluid plays a role in determining the degree of permeability of membranes.

A high concentration of calcium decreases the permeability of membranes while a low concentration (increases/decreases) permeability of membranes.

– – – – – – – – – – – – – – –

increases

118

If a membrane ruptures and there is a high calcium concentration, a new membrane will be formed immediately.

The high concentration of calcium will also decrease the *_____

_____.

– – – – – – – – – – – – – – –

permeability of the membrane

119

The major function of calcium is for the formation of bone and teeth.
It is also essential to the clotting mechanism and to normal muscle and
nerve activity.
   Name four functions of calcium in our body.
1.
2.
3.
4.

– – – – – – – – – – – – – – –

1. Maintenance of normal cell permeability
2. Formation of bone and teeth
3. Normal clotting mechanism
4. Normal muscle and nerve activity

120

The parathyroid glands, which are four small oval-shaped glands located
on the posterior thyroid gland, regulate the serum level of calcium.
These glands secrete parathyroid hormone which is responsible for the
homeostatic regulation of calcium ion in the body fluids.
   When the serum calcium level is low, the parathyroid glands will secrete
more parathyroid hormone. Explain what would happen if the serum

calcium level was high._____.

– – – – – – – – – – – – – – –

It would inhibit the secretion of the parathyroid hormone, or less
would be secreted.

121

The regulation of serum calcium is maintained by the negative feedback
system for when there is a low serum calcium, this stimulates the para-

thyroid gland to *_____, and when
there is a high serum calcium the opposite occurs, for it inhibits the
parathyroid gland in secreting hormone.

– – – – – – – – – – – – – – –

secrete parathyroid hormone

122

The negative feedback system is also responsible for the hormonal actions of most of our other body glands.

Explain the action of ADH, or antidiuretic hormone, of the posterior hypophysis or posterior pituitary gland in relation to the negative feedback system. _____

_____ .

– – – – – – – – – – – – – – – – –

When the serum solute concentration is high, then more ADH is released, which decreases the excretion of water from the renal tubules. The opposite occurs with a low serum solute concentration.

123

When there is a low serum calcium level, this tells the parathyroid gland to secrete more hormone. The parathyroid hormone will need to find calcium in order to maintain normal serum calcium. It will pull calcium from the bone, thus the bone structure will be weakened if enough is removed.

With a low serum calcium level, what would happen to the calcium in

the bone? _____ .

What would happen to the bone? _____ .

– – – – – – – – – – – – – – – – –

removal of Ca from the bone, weakened bone structure

Table 9 gives various conditions causing either a calcium deficit or excess. Study this table carefully and then proceed to the frames that follow. Refer to this table as needed.

TABLE 9 Conditions Causing Calcium Deficit and Calcium Excess

Conditions	Calcium Deficit	Calcium Excess
Inadequate protein diet Lack of calcium in diet	Inadequate protein intake inhibits the body's utilization of calcium. In our present-day society, calcium deficit is rare. Milk and milk products will increase the calcium in the body.	
Diseases, extensive infections, and burns	Calcium is pulled from extracellular fluid and is trapped in burns, diseased tissues, and infections of the great body cavity.	
Hypoparathyroidism	When the parathyroid glands are injured or destroyed, the parathyroid hormone is reduced and there is a calcium loss.	
Diarrhea	A deficit will occur when large amounts of calcium are lost in loose bowel movements.	
Hyperparathyroidism		When the parathyroid glands are overactive, calcium builds up in the extracellular fluid. This can result from tumors of the parathyroid gland, which produce excessive quantities of parathyroid hormone.
Tumors of the bones		Tumors of the bone are composed of many abnormal bone cells. The calcium from these cells is released and absorbed into the extracellular fluid.
Multiple fractures and prolonged immobilization		These conditions increase the calcium in the extracellular fluid by releasing calcium from the bones.

124

What effect does an inadequate protein diet have on calcium? _____

_____ .

– – – – – – – – – – – – – –

It inhibits the body's utilization of calcium.

125

Burns, diseases, and extensive infections can cause a calcium _____ .

Why? _____ .

– – – – – – – – – – – – – –

deficit, They pull calcium from the extracellular fluid for their repairs.

126

Hypoparathyroidism can cause a calcium _____ . How? _____

_____ .

Hyperparathyroidism can cause a calcium _____ . How? _____

_____ .

– – – – – – – – – – – – – –

deficit, Loss of the parathyroid hormone.
excess, Overproduction of the parathyroid hormone.

127

What effect does prolonged immobilization have on calcium? _____

_____ .

– – – – – – – – – – – – – –

It increases serum calcium by releasing Ca from the bones.

128

What other two conditions, not mentioned in the frames above, can

cause a calcium excess? _____

and _____.

‒ ‒ ‒ ‒ ‒ ‒ ‒ ‒ ‒ ‒ ‒ ‒ ‒ ‒

tumors of the bone and multiple fractures

129

Place a D for calcium deficit and E for calcium excess concerning the
following conditions:

( ) a. Infections
( ) b. Diseases
( ) c. Hyperparathyroidism
( ) d. Hypoparathyroidism
( ) e. Diarrhea
( ) f. Multiple fractures
( ) g. Prolonged immobilization
( ) h. Burns
( ) i. Tumor of the bone

‒ ‒ ‒ ‒ ‒ ‒ ‒ ‒ ‒ ‒ ‒ ‒ ‒

a. D	f. E
b. D	g. E
c. E	h. D
d. D	i. E
e. D	

130

Hypocalcemia means a calcium deficit in the extracellular fluid; there-

fore, another name for calcium excess would be _____.

‒ ‒ ‒ ‒ ‒ ‒ ‒ ‒ ‒ ‒ ‒ ‒ ‒

hypercalcemia

Table 10 lists signs and symptoms of hypocalcemia and hypercalcemia to the parts of the body which are affected. Study this table carefully. Know the normal range of serum calcium or the blood plasma calcium. Refer to this table as needed.

TABLE 10   Signs and Symptoms Related to Hypocalcemia and Hypercalcemia

Classes	Hypocalcemia	Hypercalcemia
Central nervous system abnormalities	Trembling of the fingers Twitching around the mouth Spasm of the voice box Irritability	Deep pain over the bony areas
Muscular abnormalities	Abdominal cramps Muscle cramps	Muscles are flabby
Blood abnormalities	Blood does not clot normally	
Urinary abnormalities		Stones composed of calcium form in the kidneys. This can result when calcium leaves the bones due to injury, immobilization, etc.
Skeletal abnormalities	Fractures occur if deficit persists due to calcium being removed from the bones	Thinning of the bones is apparent since calcium has been transferred from the bone to the extracellular fluid
Milliequivalents per liter	Below 4.5 mEq/L in blood plasma	Above 5.8 mEq/L in blood plasma

131

Indicate which of the following signs and symptoms relate to hypocalcemia:

( )   a. Cramps in the abdomen
( )   b. Abnormalities in clotting of blood
( )   c. Flabby muscles
( )   d. Stones in the kidneys
( )   e. Trembling of the fingers
( )   f. Twitching around the mouth

_ _ _ _ _ _ _ _ _ _ _ _ _ _ _

a. X   d. —
b. X   e. X
c. —   f. X

132

It may be dangerous if the amount of calcium in the extracellular fluid

is below _____ .

_ _ _ _ _ _ _ _ _ _ _ _ _ _

4.5 mEq/L

133

Indicate which of the following signs and symptoms relate to hypercalcemia:

( )   a. Cramps in the abdomen
( )   b. Abnormalities in clotting of blood
( )   c. Flabby muscles
( )   d. Stones in the urinary tract
( )   e. Deep pain over the bony areas
( )   f. Trembling of the fingers
( )   g. Thinning of the bones

_ _ _ _ _ _ _ _ _ _ _ _ _ _ _

a. —   e. X
b. —   f. —
c. X   g. X
d. X

134

It may be dangerous if the amount of calcium in the extracellular fluid

is above _____.

_ _ _ _ _ _ _ _ _ _ _ _ _ _ _

5.8 mEq/L

135

Name four symptoms related to hypocalcemia.
1.
2.
3.
4.

Name four symptoms related to hypercalcemia.
1.
2.
3.
4.

_ _ _ _ _ _ _ _ _ _ _ _ _ _ _

1. abdominal cramps
2. muscle cramps
3. abnormalities in clotting
4. trembling of fingers
5. twitching of mouth

1. flabby muscles
2. stones in the urinary tract
3. deep pain over the bony areas
4. thinning of bones

Clinical Considerations

136

Calcium acts like a sedative on the central nervous system (CNS).

Let us say you are caring for a debilitated patient, who becomes severely agitated. You notice the patient's hands trembling and mouth twitching. These symptoms might indicate a calcium (excess/deficit). What would your action be?

1.
2.

— — — — — — — — — — — — — — —

deficit
1. Check the lab report on serum calcium.
2. Notify physician of symptoms and lab report.
3. Check patient's dietary intake of Ca.

137

What food products should individuals have which are high in calcium

to prevent or correct body's calcium deficit? _____.

— — — — — — — — — — — — — —

milk and milk products

138

Normally, calcium is not required for IV therapy since there is a tremendous reservoir in the bone. However, the body needs Vitamin D for the utilization of dietary calcium.

What other essential composition of the diet is needed for calcium utilization? _____. Refer to Table 10 if needed.

— — — — — — — — — — — — — —

protein

Review—Calcium

Q. 1
Calcium is a(n) (anion/cation). Where is its highest concentration found?

_____. In what type of body fluid is its highest concentra-

tion found? _____ Calcium in the extracellular

fluid plays a role in determining the degree of * _____.

— — — — — — — — — — — — — —

cation, bone, extracellular fluid, permeability of cells

Q. 2
List at least 4 functions of calcium:
1.
2.
3.
4.

— — — — — — — — — — — — — —

1. Formation of bone and teeth
2. Normal clotting mechanism
3. Normal cell permeability
4. Normal muscle and nerve activity

Q. 3
Burns, disease, and extensive infections can cause a calcium _____.

— — — — — — — — — — — — — —

deficit

Q. 4
Hypoparathyroidism results in a calcium _____.

— — — — — — — — — — — — — —

deficit

Q. 5

Bone tumors, multiple fractures, and prolonged immobilization fre-

quently cause a calcium _____ . Why? _____

_____ .

_ _ _ _ _ _ _ _ _ _ _ _ _ _ _ _

excess, They will release calcium from bone.

Q. 6

Place D for hypocalcemia or E for hypercalcemia according to the signs
and symptoms presented:

( )   a. Cramps in the abdomen
( )   b. Abnormalities in clotting of blood
( )   c. Flabby muscles
( )   d. Stones in the urinary tract
( )   e. Deep pain over the bony areas
( )   f. Trembling of the fingers
( )   g. Twitching around the mouth
( )   h. Thinning of the bones
( )   i. 8.6 mEq/L in blood plasma
( )   j. 6.4 mEq/L in blood plasma
( )   k. 3.8 mEq/L in blood plasma

_ _ _ _ _ _ _ _ _ _ _ _ _ _ _ _

a. D   g. D
b. D   h. E
c. E   i. E
d. E   j. E
e. E   k. D
f. D

MAGNESIUM

139

Magnesium is a(n) (cation/anion). Its highest concentration is found in
what type of fluid? _____. Refer to Table 3 if needed.

— — — — — — — — — — — — — —

cation, intracellular

140

What other cation has its highest concentration in the intracellular
fluid? _____.

— — — — — — — — — — — — — —

potassium

141

Magnesium is widely distributed throughout the body. Most of the body
magnesium is in the bone. What other ion is found plentifully in the
bone? _____.

— — — — — — — — — — — — — —

calcium

142

Magnesium has a higher concentration in the cerebrospinal fluid, also
known as spinal fluid, than in the blood plasma, The serum concentration of magnesium is _____. Refer to Table 3 if needed.

— — — — — — — — — — — — — —

2 mEq/L; actually the range is 1.7 – 2.3 mEq/L

143

Magnesium plays an important role in enzyme activity. An enzyme is a catalyst capable of inducing chemical changes in other substances. Magnesium acts as a coenzyme in the metabolism of carbohydrates and protein.

Magnesium is also involved in maintaining neuromuscular stability.

What other ion has this similar function? _____.

_ _ _ _ _ _ _ _ _ _ _ _ _ _ _

calcium

144

Hypomagnesemia means a low serum concentration of magnesium.

Therefore, hypermagnesemia would mean _____.

_ _ _ _ _ _ _ _ _ _ _ _ _ _ _

high serum concentration of magnesium

Table 11 lists the conditions causing hypomagnesemia and hypermagnesemia, each with an explanation. Study this table carefully and refer to the table as needed. Refer to the glossary for unknown words.

TABLE 11 Conditions Causing Magnesium Deficit and Magnesium Excess

Conditions	Magnesium Deficit— Hypomagnesemia	Magnesium Excess— Hypermagnesemia
Prolonged inadequate intake of food	Magnesium is found in various foods, particularly green vegetables. Severe anorexia with minimum to no food could cause a deficit.	
	Continuous intravenous therapy without magnesium supplement.	
Chronic alcoholism, malnutrition	An inadequate intake of food over a long period of time causes a deficit.	
Prolonged diuresis	It will cause a magnesium loss via urine.	
Severe dehydration (especially patients with diabetic acidosis)	Loss of magnesium from the intracellular fluid and from the body.	
Renal insufficiency, uremia		Retention of magnesium in the blood plasma.
Epsom salts as a laxative (magnesium sulfate is the same as epsom salts)		It will increase the concentration of magnesium in the blood plasma if given in excessive amounts over a prolonged period of time.

145

Magnesium is found in various foods, but a prolonged inadequate intake could cause a *_____.

— — — — — — — — — — — — — — —

magnesium deficit

146

Chronic alcoholism and malnutrition are conditions that can cause magnesium_____.

_ _ _ _ _ _ _ _ _ _ _ _ _ _ _ _

deficit

147

Two other situations which could cause hypomagnesemia are

*_____and_____.

_ _ _ _ _ _ _ _ _ _ _ _ _ _ _ _

continuous IV therapy, prolonged diuresis, or severe dehydration

---

148

Prolonged renal insufficiency could cause a *_____.

_ _ _ _ _ _ _ _ _ _ _ _ _ _ _ _

magnesium excess

149

Place a D for magnesium deficit and an E for magnesium excess concerning the following:

( ) a. Renal insufficiency
( ) b. Prolonged diuresis
( ) c. The use of epsom salts as a laxative
( ) d. Chronic alcoholism
( ) e. Malnutrition
( ) f. Prolonged inadequate intake
( ) g. Severe dehydration—diabetic acidosis

_ _ _ _ _ _ _ _ _ _ _ _ _ _ _

a. E   e. D
b. D   f. D
c. E   g. D
d. D

Study Table 12 carefully and refer to this table as needed.

TABLE 12 Signs and Symptoms Related to Hypomagnesemia and Hypermagnesemia

Classes	Hypomagnesemia	Hypermagnesemia
Central nervous system abnormalities	Hyperirritability; bizarre, involuntary muscular activity appears as tremors. Twitching of the face. Spasticity Convulsion	CNS depression. Decreased respiration. Inhibition of neuromuscular transmission. Lethargy Coma
Cardiac abnormalities	Cardiac arrhythmia Vasodilatation resulting in a blood pressure decrease.	Bradycardia

150

Hyperirritability, bizarre muscular activity, twitching of the face, and

convulsions are signs and symptoms of _____ .

— — — — — — — — — — — — — — —

hypomagnesemia

151

Give a name of another ion which has similar symptoms of "hypo"

effect._____ .

— — — — — — — — — — — — —

calcium

152

Central nervous system depression, inhibited neuromuscular trans-
mission, decreased respiration and lethargic state are signs and symptoms

of _____ .

— — — — — — — — — — — — —

hypermagnesemia

153

Match the following cardiac signs and symptoms relating to a magnesium deficit or excess:

_____ 1. Cardiac arrhythmia	a. Hypomagnesemia	
_____ 2. Vasodilatation resulting in a blood pressure decrease	b. Hypermagnesemia	
_____ 3. Bradycardia		

- - - - - - - - - - - - - - - -

1. a
2. a
3. b

154

Place D for hypomagnesemia and E for hypermagnesemia concerning the following signs and symptoms:

( ) a. Hyperirritability
( ) b. CNS depression
( ) c. Lethargy
( ) d. Bizarre muscular activity resembling tremors
( ) e. Twitching of the face
( ) f. Convulsion
( ) g. Decreased respiration
( ) h. Bradycardia
( ) i. Cardiac arrhythmia
( ) j. Vasodilatation

- - - - - - - - - - - - - - -

a. D	f. D
b. E	g. E
c. E	h. E
d. D	i. D
e. D	j. D

155

Calcium has an antagonistic effect upon an increase in serum magnesium. Though calcium antagonizes magnesium overdosage, both low magnesium and low calcium cause similar CNS abnormalities such as

*_____.

– – – – – – – – – – – – – – – – –

twitchings, convulsions, tremors, irritability

156

For the management of hypomagnesemia, establishing adequate renal flow is first indicated and then rehydration and administering the ion,

_____.

   Why would renal sufficiency be needed when administering magnesium? _____

_____.

– – – – – – – – – – – – – –

magnesium
It is necessary in excreting excess magnesium if too much magnesium was administered.

Clinical Considerations

157

It has been reported that magnesium will slow the rate of auricular flutter, affecting the atrium of the heart. It also has been reported that it will restore normal sinus rhythm, which is normal heart rhythm, and will diminish arrhythmias produced by digitalis and potassium. How does a potassium deficit affect the heart muscle? _____

_____.

– – – – – – – – – – – – – – –

It will cause arrhythmias and eventually death.

158

Name two cations necessary for normal heart rhythm. _____

and _____.

- - - - - - - - - - - - - - - - -

potassium and magnesium

159

In diabetic acidosis, magnesium will leave the cells. When insulin and dextrose are given intravenously, magnesium will return to the cells.

If the diabetic condition is corrected too fast, then (hypomagnesemia/ hypermagnesemia) would occur. Why? _____

_____.

- - - - - - - - - - - - - - -

hypomagnesemia, Mg is drawn from the extracellular fluid too fast.

Review—Magnesium

Q. 1

Name four cations that have been discussed so far in this chapter.

_____ , _____ ,

_____ , and _____ .

- - - - - - - - - - - - - - -

potassium, sodium, calcium, and magnesium

Q. 2

Name the two cations that have their highest concentration in the intra-

cellular fluid. _____ and _____.

- - - - - - - - - - - - - - -

potassium and magnesium

Q. 3

A chronic alcoholic condition resulting in malnutrition could cause

_____. Why? _____.

_ _ _ _ _ _ _ _ _ _ _ _ _ _

hypomagnesemia, Inadequate food intake

Q. 4

Prolonged diuresis can cause a magnesium _____. Why _____

_____.

_ _ _ _ _ _ _ _ _ _ _ _ _ _

deficit, Large amount would be excreted via kidneys.

Q. 5

Epsom salts given in large amounts or over prolonged period of time

will cause_____.

_ _ _ _ _ _ _ _ _ _ _ _ _ _

hypermagnesemia

Q. 6

Renal insufficiency can cause (hypomagnesemia/hypermagnesemia).

_ _ _ _ _ _ _ _ _ _ _ _ _ _

hypermagnesemia

Q. 7

Three signs and symptoms of hypomagnesemia are:
1.
2.
3.

_ _ _ _ _ _ _ _ _ _ _ _ _ _

Convulsions, hyperirritability, bizarre muscular activity, twitching of
the face, cardiac arrhythmia

Q. 8
Three signs and symptoms of hypermagnesemia are:
1.
2.
3.

- - - - - - - - - - - - - -

CNS depression, decreased respiration, lethargic, bradycardia

Q. 9
The two electrolytes that have similar CNS abnormalities when in low

concentration are _____ and _____ .

- - - - - - - - - - - - - -

calcium and magnesium

CHLORIDE

160
Chloride is a(n) (anion/cation).
The chloride ion frequently appears in combination with the sodium ion. Which fluid has the greatest concentration of chloride—intracellular or extracellular?

- - - - - - - - - - - - - -

anion, extracellular fluid

161
Refer to Table 3 as needed. The concentration of chloride in plasma is

_____ , and in intracellular fluid is _____ .

- - - - - - - - - - - - - -

104 mEq/L, 1 mEq/L, though the range could be 1 – 4 mEq/L

162
Highest chloride concentration is in the cerebrospinal fluid, 125 mEq/l.

The greatest concentration of chloride is in *_____ .

- - - - - - - - - - - - - -

cerebrospinal fluid

163

If there is a deficiency of chloride, this can lead to a deficiency of potassium and vice versa. Usually chloride losses will follow those of sodium. Most chloride ingestion is in combination with sodium.

The chloride ion is mostly found in combination with a

_____ ion.

What is the name for sodium chloride? _____.

_ _ _ _ _ _ _ _ _ _ _ _ _ _ _

sodium, salt

164

For every sodium ion absorbed from the renal tubules, a chloride or bicarbonate ion is also absorbed, thus the loss of sodium and chloride in proportion can differ.

The organ responsible for electrolyte homeostasis by the extretion

and absorption of ions is the _____.

_ _ _ _ _ _ _ _ _ _ _ _ _ _ _

kidneys

165

Chloride has an important place in the acid-base balance and body water balance, which will be discussed in detail in Chapter 3.

When sodium is retained, chloride frequently is retained, causing an increase in water retention. Can you explain the reason why water would

be retained? _____.

_ _ _ _ _ _ _ _ _ _ _ _ _ _

Sodium holds water, which causes edema.

166

Chloride plays a part in oxygen and carbon dioxide exchange in the red blood cells. When the red blood cells are oxygenated, chloride travels from the red blood cells to the plasma and bicarbonate leaves the plasma to the red blood cells. This is called the chloride shift.

This chloride shift is necessary in maintaining _____

_____ .

_ _ _ _ _ _ _ _ _ _ _ _ _ _ _

homeostasis or equilibrium

167

When chloride leaves the red blood cells in the arterial blood as the oxygen enters, the plasma concentration would be (lowered/elevated). In the venous blood the chloride plasma concentration would be

_____ .

_ _ _ _ _ _ _ _ _ _ _ _ _ _ _

elevated, lowered

168

Chloride also combines with the hydrogen ion in the stomach to produce the acidity of the gastric juice.

Name at least three functions of the chloride ion.

1.
2.
3.

_ _ _ _ _ _ _ _ _ _ _ _ _ _

1. Acid-base balance
2. Body water balance
3. Acidity of the gastric juice

169

The name for chloride deficit is hypochloremia; so therefore, the name

for chloride excess is _____ .

– – – – – – – – – – – – – –

hyperchloremia

Table 13 lists conditions causing either a chloride deficit or excess. Study the table carefully and refer to the glossary for unknown words. Refer back to this table as needed.

TABLE 13   Conditions Causing Chloride Deficit and Chloride Excess

Condition	Chloride Deficit–Hypochloremia	Chloride Excess–Hyperchloremia
Continuous vomiting, gastric suction	The acidity of the gastric juice is composed of hydrogen and chloride. A loss of gastric juice via vomiting or through gastric suction will cause a chloride deficit.	
Diarrhea	A loss of the gastrointestinal salty secretions will cause a chloride deficit.	
Loss of potassium	A loss of potassium is accompanied by loss of chloride.	
Prolonged use of mercurial diuretics	Mercurial diuretics interfere with the absorption of chloride ions from the renal tubules.	
Excessive sweating	Chloride, combined with sodium, is lost via skin. This can result from increased temperature, fever, or muscular exercise.	
Excessive adrenocortical hormone production		Excessive adrenal cortical hormone will cause an excess of sodium in the body. Sodium combines with chloride increasing the chloride ion in the blood serum.

170

Name the two ions that cause the acidity of the gastric juice.

_____ and _____.

What gastrointestinal conditions can cause a chloride deficit?

_____ , _____ , and _____.

– – – – – – – – – – – – – – –

hydrogen and chloride
1. continuous vomiting
2. gastric suction
3. diarrhea

171

The loss of chloride is frequently accompanied by the loss of what other

ion? _____.

– – – – – – – – – – – – – – –

potassium. Yes, sodium could be an answer too.

172

Chloride is frequently combined with the _____ ion.

Name at least two causes for sodium and chloride loss or salt loss

through the skin. _____ and _____.

– – – – – – – – – – – – – –

sodium. Yes, hydrogen could be an answer too.
fever, muscular exercise, and increased environmental temperature

173

What is the effect of prolonged use of mercurial diuretics? _____

_____.

– – – – – – – – – – – – – –

They interfere with the absorption of chloride ions from kidneys.

174

Excessive adrenocortical hormone can cause a sodium excess and a

chloride _____? Why _____

_____.

– – – – – – – – – – – – – –

excess
Sodium combines with chloride increasing the chloride ion in the blood serum.

175

Place D for chloride deficit and E for chloride excess concerning the following:
( ) a. Vomiting
( ) b. Diarrhea
( ) c. Gastric suction
( ) d. Excessive adrenocortical hormone production
( ) e. Sweating
( ) f. Mercurial diuretics
( ) g. Potassium loss

– – – – – – – – – – – – – –

a. D   e. D
b. D   f. D
c. D   g. D
d. E

176

Another name for chloride deficit is _____ and an-

other name for chloride excess is _____.

– – – – – – – – – – – – – –

hypochloremia, hyperchloremia

Review—Chloride

Q. 1

Chloride is a(an) (cation/anion), and its greatest concentration is found in the *_____.

_ _ _ _ _ _ _ _ _ _ _ _ _ _ _ _

anion, cerebrospinal fluid

Q. 2

Practically all of the chloride ingested is in combination with _____.

_ _ _ _ _ _ _ _ _ _ _ _ _ _ _

sodium

Q. 3

Chloride combines with the hydrogen ion to produce *_____

_____.

_ _ _ _ _ _ _ _ _ _ _ _ _ _ _

the acidity of the gastric juice

Q. 4

Vomiting, gastric suction, and diarrhea can result in_____.

_ _ _ _ _ _ _ _ _ _ _ _ _ _ _

hypochloremia

Q. 5

Prolonged use of mercurial diuretics and sweating can cause

_____.

_ _ _ _ _ _ _ _ _ _ _ _ _ _ _

hypochloremia

Q. 6

As adrenal cortical hormone production increases, what happens to the chloride blood serum? _____

_____.

_ _ _ _ _ _ _ _ _ _ _ _ _ _ _

It will increase due to the increase of the sodium ion as the result of the AC hormone.

Q. 7

Explain the chloride shift. _____

_____.

_ _ _ _ _ _ _ _ _ _ _ _ _ _

When chloride leaves the red blood cells, bicarbonate enters and vice versa.

Q. 8

Three main functions of the chloride ion are:
1.
2.
3.

_ _ _ _ _ _ _ _ _ _ _ _ _ _

1. Acid-base balance
2. Body water balance
3. Acidity of the gastric juice

Table 14 lists the foods by name which are rich in electrolyte content, according to classes of food. Where no foods are listed, all of the foods in that class are low in that electrolyte. You will, no doubt, need to refer back to this table. However, it is important that you memorize the foods rich in potassium. Due to the potassium deficit which frequently can occur, you need to remember the foods rich in this electrolyte. Again, study this table very carefully and then proceed to the frames that follow.

TABLE 14 Foods Rich in Potassium, Sodium, Calcium, Magnesium, and Chloride (mg/100 grams)

Classes	Potassium	Sodium	Calcium	Magnesium	Chloride
Daily requirements	3 – 4 grams	2 – 4 grams	800 mg	300 mg	3 – 9 grams
Beverages	Cocoa, Coca Cola, Coffee, Wines	Pepsi-Cola, Tea, Coffee, decaffinated		Cocoa	
Fruit and fruit juices	Citrus fruits Oranges, Grapefruit Juices Grapefruit (canned) Orange (canned) Prune (canned) Tomato (canned) Fruits Apricots (dry), Bananas, Cantaloupe, Dates, Raisins (dry), Watermelon			Average	High only in Dates and Bananas
Bread products	Average to low amount	White bread, Soda crackers, Wheat flakes		Cereals with oats	
Dairy products	Milk, Buttermilk	Butter, Cheese, Margarine	Milk, Cheese,	Milk, average	Cheese, Milk
Nuts	Almonds, Brazil nuts, Cashews, Peanuts	Low, except if salted	Brazil nuts, moderate	Almonds, Brazil nuts, Peanuts, Walnuts	

Vegetables	Baked beans, Carrots (raw), Celery (raw), Dandelion greens, Lima beans (canned), Mustard greens, Tomatoes, Spinach NOTE: Nearly all vegetables are rich in potassium when raw but K will be lost if water used in cooking is discarded.		Average to low; Celery, high average	Baked beans, Kale, Mustard and Turnip Greens, Broccoli	Spinach, Celery
Meat, fish, and poultry	Average	Corned beef, Bacon, Ham, Crab, Tuna fish; Low in poultry	Salmon, Meats	Bacon, Shrimp, Low in poultry, Low in meats, Egg, average	Eggs, Crabs, Fish, average, Turkey
Miscellaneous	Catsup, average, Spices, Potato chips	Catsup, Mayonnaise, Potato chips, average, Pickles, dill, Olives, Mustard, paste, Worcestershire sauce, Celery salt	Molasses	Chocolate and chocolate bars, Chocolate syrup, Molasses, Table salt	

177

What classes of food are rich in potassium? _____ ,

_____ , _____ , _____ ,

and _____ .

_ _ _ _ _ _ _ _ _ _ _ _ _ _ _

dairy products, nuts, vegetables, fruits, and fruit juice

178

Many American people consume 5 to 15 grams of sodium chloride per day, which is more than needed.

The daily requirement for sodium per day is _____ .

The daily requirement for potassium per day is _____ .

_ _ _ _ _ _ _ _ _ _ _ _ _ _ _

2 to 4 grams, 3 to 4 grams

179

Name the classes of food that are rich in sodium. _____ ,

_____ , _____ , and _____ .

_ _ _ _ _ _ _ _ _ _ _ _ _ _ _

bread products, dairy products, meat, and fish

180

The classes of food that are rich in calcium are:
1.
2.
3.
4.

_ _ _ _ _ _ _ _ _ _ _ _ _ _ _

1. bread products
2. dairy products
3. fish (salmon)
4. meat

181

The classes of food that are rich in magnesium are oats from bread products, shrimp from fish, and _____ .

— — — — — — — — — — — — — —

nuts

182

Give examples of foods that are rich in chloride:
1. fruits: _____ .
2. meats: _____ and _____ .
3. fish: _____ .
4. dairy products: _____ and _____ .
5. vegetables: _____ and _____ .

— — — — — — — — — — — — — —

1. dates or bananas
2. turkey and egg
3. crab
4. cheese and milk
5. spinach and celery

183

Check the classes of foods rich in K (potassium), Na (sodium), Ca (calcium, Mg (magnesium), and Cl (chloride).

```
      K    Na   Ca   Mg   Cl
a. (  ) (  ) (  ) (  ) (  ) Fruits
b. (  ) (  ) (  ) (  ) (  ) Fruit juices
c. (  ) (  ) (  ) (  ) (  ) Bread products
d. (  ) (  ) (  ) (  ) (  ) Dairy products
e. (  ) (  ) (  ) (  ) (  ) Nuts
f. (  ) (  ) (  ) (  ) (  ) Vegetables
g. (  ) (  ) (  ) (  ) (  ) Meat
h. (  ) (  ) (  ) (  ) (  ) Fish
```

_ _ _ _ _ _ _ _ _ _ _ _ _ _ _ _

a. K, Cl (dates)
b. K
c. Na, Mg (oats)
d. K, Na, Ca, Cl
e. K, Mg
f. K, Ca, Cl
g. Na, Cl, Ca
h. Na, Ca, Mg, Cl

184

Place K for potassium and Na for sodium relating to the classes of food in which these electrolytes are found in great concentration:

a. _____ Fruits

b. _____ Fruit juices

c. _____ Meat

d. _____ Fish

e. _____ Bread

f. _____ Shelled nuts

g. _____ Dairy products

h. _____ Vegetables

- - - - - - - - - - - - - - - -

a. K	e. Na
b. K	f. K
c. Na	g. K, Na
d. Na	h. K

SUMMARY QUESTIONS

The questions in this summary will test your comprehensive ability and knowledge of the material found in Chapter 2. If the answer is unknown, refer to the section in the chapter for further understanding and clarification.

1. What are electrolytes? What are ions? Differentiate between cations and anions.
2. Too much serum potassium is known as _____ and too little serum potassium is known as _____.
   What is the "normal" range for serum potassium?
3. Give at least 6 clinical conditions which can cause a potassium deficit, and at least 2 clinical conditions which can cause a potassium excess.
4. List the common symptoms found with hyperkalemia and hypokalemia.

5. What classes of food are rich in potassium?
6. Sodium is the main (anion/cation) found in the extracellular fluid. The "normal" serum concentration range of sodium is

   _____ .

7. Explain how the kidneys, hypophysis, and the adrenal cortical hormones affect serum sodium.
8. Give at least 6 clinical conditions which can cause a sodium deficit and at least 3 clinical conditions which can cause a sodium excess.
9. What are the most common symptoms of hyponatremia and hypernatremia?
10. What classes of foods are rich in sodium?
11. List the four functions of calcium.
12. Give at least 4 clinical conditions which can cause a calcium deficit and at least 3 clinical conditions which can cause a calcium excess.
13. What are the common symptoms of hypocalcemia and hypercalcemia?
14. What classes of foods are rich in calcium?
15. Give at least 4 clinical conditions which can cause a magnesium deficit and at least 2 clinical conditions which can cause a magnesium excess.
16. The two electrolytes which have similar CNS abnormalities when their

    serum levels are low are _____ and _____ .
17. What are the symptoms of hypomagnesemia and hypermagnesemia?
18. Practically all of the chloride ingested is in combination with

    _____ .

19. What are the three main functions of the chloride ion?
20. What clinical conditions can cause a hypochloremic state or a hyperchloremic state?

References

Abbott Laboratories, *Fluid and Electrolytes*, North Chicago, Illinois, 1968, pp. 15–21.
Baxter Laboratories, *The Fundamentals of Body Water & Electrolytes*, Morton Grove, Illinois, 1967, pp. 9–21.
Burgess, Richard, "Fluids and Electrolytes," *American Journal of Nursing*, LXV (October, 1965), pp. 92–93.

Cooper, Lenna, *et al.*, *Nutrition in Health and Disease*, 14th ed., Philadelphia, Pa.: J. B. Lippincott Co., 1963, pp. 530–535.

Crowell, Caleb E., ed., "Programmed Instruction—Potassium Imbalance," *American Journal of Nursing*, LXVII (February, 1967), pp. 343–366.

De Veber, George A., "Fluid and Electrolyte Problems in the Post-Operative Period," *The Nursing Clinics of North America*, I (June, 1966), pp. 280–283.

Dutcher, Isabel E. and Sandra B. Fielo, *Water and Electrolytes*, New York: The Macmillan Co., 1967, pp. 19–28, p. 33.

Metheney, Norma M. and William D. Snively, *Nurses' Handbook of Fluid Balance*, Philadelphia, Pa.: J. B. Lippincott Co., 1967, pp. 22–27.

Shafer, Kathleeen N., *et al.*, *Medical-Surgical Nursing*, 3rd ed., St. Louis, Mo.: The C. V. Mosby Co., 1964, pp. 60–62.

Snively, William D., *Sea Within*, Philadelphia, Pa.: J. B. Lippincott Co., 1960, pp. 62–81.

Statland, Harry, *Fluid and Electrolytes in Practice*, 3rd ed., Philadelphia, Pa.: J. B. Lippincott Co., 1963, pp. 98–126.

Turner, Dorothea, *Handbook of Diet Therapy*, 3rd ed., Illinois: University of Chicago Press, 1959, appendix.

Wilson, Eva D. *et al.*, *Principles of Nutrition*, 2nd ed., New York: John Wiley and Sons, Inc., 1966, p. 184.

Wohl, Michael G. and Robert S. Goodhart, *Modern Nutrition in Health and Disease*, 3rd ed., Philadelphia, Pa.: Lea and Febiger, 1964, pp. 481–483.

Chapter 3

Acid-Base Balance and Imbalance

BEHAVIORAL OBJECTIVES

Upon completion of this chapter, the student will be able to:

Explain the influence of the hydrogen ion on body fluids.

Give the pH ranges for acidosis and alkalosis.

Explain the four regulatory mechanisms for pH control and how the regulatory mechanisms can maintain acid-base balance.

Explain how various clinical conditions can cause metabolic acidosis and alkalosis and respiratory acidosis and alkalosis.

Observe clinically for symptoms of metabolic acidosis and alkalosis and respiratory acidosis and alkalosis. (You may need some assistance in recognizing all these symptoms.)

Explain the body's defense action and the clinical management for acid-base balance and be able to apply this with assistance to various clinical situations.

INTRODUCTION

Our body fluid must maintain a balance between acidity and alkalinity in order for life to be maintained. "Acid" comes from the Latin word meaning "sharp" and acid is frequently referred to as being sour. On the other hand, alkaline is referred to as being sweet. According to Bronsted-Lowry concept of acids and bases, an "acid is any molecule or ion which can donate a proton to any other substance, whereas a base is any molecule or ion which can accept a proton" (Best and Taylor, p. 596). Also the more readily an acid gives up its protons the stronger it is as an acid. Acids and bases are not synonymous with anions and cations.

Other theories state that the concentration of the hydrogen ion (plus or minus) will determine either the acidity or the alkalinity of a solution. The amount of ionized hydrogen in the extracellular fluid is extremly small; around .0000001 gram per liter. Instead of using this cumbersome figure, the symbol pH is used which stands for the negative logarithm (exponent) of the hydrogen ion concentration. Mathematically, it would be expressed as 10^{-7}, the base being 10 and the power minus 7 being the logarithm of the number. The minus sign is dropped and the symbol used to designate the hydrogen ion concentration would then be pH 7. As the hydrogen ion concentration rises (in solution) the pH value falls, indicating a decreased negative logarithm of the hydrogen ion concentration, thus indicating acidity. As the hydrogen concentration falls, the pH rises, thus indicating alkalinity.

The hydroxyl ions (OH^-) are base ions and when in excess will cause the alkalinity of the solution. A solution of pH 7 is neutral since at that concentration the number of hydrogen ions (H^+) is exactly balanced by the number of hydroxyl ions (OH^-).

The above information will help you in the basic understanding of acidity and alkalinity. This will aid in understanding the material presented in this chapter: the regulatory mechanism for pH control; the clinical conditions causing metabolic acidosis and alkalosis and respiratory acidosis and alkalosis; symptoms and clinical management of acidosis and alkalosis; and clinical considerations.

Refer to the Introduction as needed to answer the first six frames.

1

According to Bronsted-Lowry concept of acids and bases, an acid is a proton (donor/acceptor) and a base is a proton (donor/acceptor).

— — — — — — — — — — — — — — —

donor, acceptor

2

For our purposes, we shall consider that the acidity or alkalinity of a solution depends upon the concentration of the *_____.

An increase in concentration of the hydrogen ions makes a solution more _____ and a decrease makes it more _____.

— — — — — — — — — — — — — —

hydrogen ions, acid, alkaline

3

Explain the meaning of the pH symbol. _____

_____.

   What effect does this have on a solution? _____

_____.

— — — — — — — — — — — — — —

negative logarithm of the hydrogen ion
It determines the acidity or alkalinity of a solution.

4

As the hydrogen ion concentration increases, the pH value _____

_____. What does that indicate? _____

_____.

   As the hydrogen ion concentration falls, the pH value _____

_____. What does that indicate? _____

_____.

— — — — — — — — — — — — — —

falls or decreases.
Indicates acidity—a decrease in the negative logarithm of the H ion.
rises or increases.
Indicates alkalinity—an increase in the negative logarithm.

5

A pH of 7 represents ten times the number of hydrogen ions as does a

pH of 8. Which of the two pH's would be considered alkaline? _____.

— — — — — — — — — — — — — —

pH 8

6

A solution at pH 7 is neutral. Why? _____

_____ .

The symbol for the hydrogen ion is _____ and for the hydroxyl ion

is _____. A hydroxyl ion and CO_2 would yield _____

_____ .

_ _ _ _ _ _ _ _ _ _ _ _ _ _ _ _

The number of hydrogen ions is balanced by the number of hydroxyl ions.

H^+, OH^-, HCO_3 known as bicarbonate

7

The pH of extracellular fluid in health is maintained at a level between 7.35 and 7.45. The body fluid is slightly (acid/alkaline).

_ _ _ _ _ _ _ _ _ _ _ _ _ _ _ _

alkaline

8

With a pH higher than this range (7.35–7.45), the body is considered to to be in a state of alkalosis.

What do you think we would call a state where the pH is below 7.35?

_____ .

_ _ _ _ _ _ _ _ _ _ _ _ _ _ _ _

acidosis

9

The pH norm of blood serum is 7.4; a variation of four tenths of a pH unit in either direction can be fatal.

In a healthy individual, the pH range of blood serum is _____ .

_ _ _ _ _ _ _ _ _ _ _ _ _ _ _ _

7.35 – 7.45

10
Whether a substance is acid, neutral, or alkaline depends upon the number of _____ ions present in a given weight or volume.

— — — — — — — — — — — — —

hydrogen

11
When the number of hydrogen ions increases in the body fluid, the body fluid becomes _____.
    When the number of hydrogen ions decreases, the body fluid becomes

_____.

— — — — — — — — — — — — —

acid, alkaline

12
In health, there are 1 1/3 mEq/L of acid to each 27 mEq of alkali for each liter of extracellular fluid, which represents a ratio of 1 part of acid to 20 parts of alkali.
    Why do think the measurement of acid and alkali are based on the extracellular fluid? _____.
    If the ratio of 1–20 is maintained, the patient is said to be in acid-base (balance/imbalance).

— — — — — — — — — — — — —

more available for analysis, balance

    Diagram 7 demonstrates by the arrow that the body is in acid-base balance when there is 1 part acid to 20 parts alkali. The pH would be 7.4. If the arrow tilts left due to an alkali deficit or acid excess, then acidosis would occur, and if the arrow tilts right due to an alkali excess or acid deficit, then alkalosis would occur. H_2CO_3 is carbonic acid.
    Study this diagram carefully. Know what will happen when the arrow tilts either left or right. Refer to the diagram when needed.

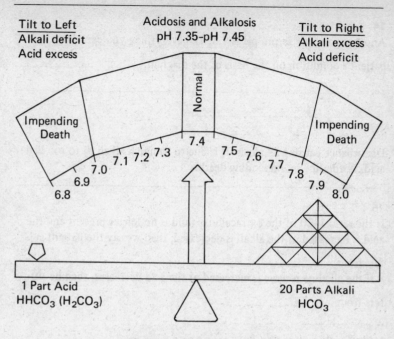

Diagram 7. Acidosis and alkalosis.

13

Your patient's serum pH is 7.1. Tell everything you can about the patient's condition on the basis of Diagram 7. _____

_____

_____ .

— — — — — — — — — — — — — — —

An acidotic condition has occurred. It is due to too much acid or too little alkali in the extracellular fluid.

14

Another patient's serum pH is 7.8. Tell everything you can about the patient's condition on the basis of the diagram. _____

_____

_____.

_ _ _ _ _ _ _ _ _ _ _ _ _ _ _ _

Disturbance causing alkalosis. It is due to to much alkali or to too little acid. Borderline on impending death.

15

If the ratio 1–20 of the extracellular fluid is no longer present and the acid is increased or the alkali is decreased, then we say the patient suffers from _____.

If the alkaline reserve is increased or the acid decreases, then he suffers from _____.

_ _ _ _ _ _ _ _ _ _ _ _ _ _ _

acidosis, alkalosis

16

Which of the following will occur as the result of ACIDOSIS? The

The balance is tilted _____.
    There is/are:
(   ) a. alkali deficit
(   ) b. alkali excess
(   ) c. acid deficit
(   ) d. acid excess
    The pH is below _____.

_ _ _ _ _ _ _ _ _ _ _ _ _ _

left
a. X   c. —
b. —   d. X
7.35

17

Which of the following will occur as the result of ALKALOSIS? The

balance is tilted _____.
   There is/are:
( ) a. alkali deficit
( ) b. alkali excess
( ) c. acid deficit
( ) d. acid excess
   The pH is above _____.

_ _ _ _ _ _ _ _ _ _ _ _ _ _ _

right
a. −   c. X
b. X   d. −
7.45

18

The body is provided with several mechanisms for controlling its pH
even though considerable amounts of acid or alkali enter the body.

   The pH of blood serum in a healthy individual is between _____.

   The pH norm of blood serum is _____.

_ _ _ _ _ _ _ _ _ _ _ _ _ _

7.35 and 7.45, 7.4

19

The main regulatory mechanisms for pH control are:
1. The Buffer System
2. The Ion Exchange
3. The Respiratory Regulation
4. The Renal Regulation
    Name the four regulatory mechanisms for pH control.
1.
2.
3.
4.

— — — — — — — — — — — — — — —

1. Buffer System
2. Ion Exchange
3. Respiratory Regulation
4. Renal Regulation

Table 15 is divided into three parts. Part A explains the Buffer System. Buffers maintain the acid-base balance of body fluids by protecting the fluid against changes in pH. They act like chemical sponges for they can soak up surplus hydrogen ions or can release them.

There are four main examples of the buffer system; bicarbonate-carbonic acid, phosphate, hemoglobin-oxyhemoglobin, and protein. These are presented in Table 15A along with their intervention—their action as a buffer—and the rationale—the reason for their action. The bicarbonate-carbonic acid buffer system is more readily available and, therefore, it is the principal buffer system of the body. Strong acids when added will combine with the bicarbonate ion to form carbonic acid, which is a weak acid. This prevents the fluids from becoming strongly acid.

Remember that bicarbonates and phosphates are anions. Refer to the glossary for unknown words. Study the table carefully and refer to it as needed.

TABLE 15 The Regulatory Mechanism for pH Control. A. Buffer Systems

Regulatory Mechanism	Intervention	Rationale
a. Bicarbonate– carbonic acid buffer system. (Principal buffer system of the body.)	Acids combine with the bicarbonates in the blood to form neutral salts (bicarbonate salt) and carbonic acid (weak acid). Carbonic acid (H_2CO_3) is a weak and unstable acid changing to water and carbon dioxide in fluid ($H_2CO_3 \rightleftharpoons H_2O + CO_2$).	When a strong acid enters the body, the H^+ ions are picked up by the bicarbonate changing it to carbonic acid ($HCO_3 + H^+ \rightarrow H_2CO_3$). The weak acid ionizes more poorly than the strong acid and it does not release H^+ ions as readily as a strong acid. If a base is added to the system it is neutralized by the carbonic acid to form water and bicarbonate ($H_2CO_3 + OH \rightarrow H_2O + HCO_3$).
b. Phosphate buffer system	The phosphate buffer system increases the amount of sodium bicarbonate ($NaHCO_3$) in the extracellular fluids, making extracellular fluids more alkaline. The H^+ ion is excreted as NaH_2PO_4 and the Na and bicarbonate ions combine.	Excess hydrogen ions combine in the renal tubules with Na_2HPO_4 (disodium phosphate) forming NaH_2PO_4 ($H^+ + Na_2HPO_4 \rightarrow Na + NaH_2PO_4$); the Na ion is reabsorbed (Na + HCO_3 $\rightarrow NaHCO_3$) and the hydrogen ion is passed into the urine.
c. Hemoglobin– Oxyhemoglobin buffer system	Maintains the same pH level in venous blood and as in arterial blood.	The venous blood has a higher CO_2 content and bicarbonate ion concentration than the arterial blood, but the pH is the same since the oxyhemoglobin (acid) in the erythrocyte has taken over some anion function which was provided by the excess bicarbonate in the venous plasma.

TABLE 15 (continued)

Regulatory Mechanism	Intervention	Rationale
d. Protein buffer-system	Proteins can exist in the form of acids (H protein) or alkaline salts (B protein) and in this way are able to bind or release excess hydrogen as required.	Proteins are amphoteric carrying both an acidic and basic charge.

20

Explain the purpose of the buffer system. _____

_____ .

How do the buffer systems accomplish their purpose? _____

_____ .

– – – – – – – – – – – – – – – –

They protect fluids against changes in pH.
They soak up surplus H^+ ions and release them as needed.

21

What is the principal buffer system of the body? _____

_____ .

– – – – – – – – – – – – – – – –

bicarbonate–carbonic acid buffer system.

22

When acid enters the body, the hydrogen ion is picked up by the bi-

carbonate, changing it to *_____ .

A base added to the body is neutralized by carbonic acid to form

_____ and _____ .

– – – – – – – – – – – – – –

carbonic acid, water and bicarbonate

23

The phosphate buffer system maintains the acid-base balance by combin-

ing the excess hydrogen ions with sodium salts, forming _____

_____. The H$^+$ is excreted in _____.

This system (excretes/retains) excess acid in the body.

— — — — — — — — — — — — —

NaH_2HPO_4 or disodium phosphate, urine, excretes

24

What is the function of the hemoglobin–oxyhemoglobin buffer system?

_____

_____.

— — — — — — — — — — — — — —

It maintains the same pH level in the venous blood and arterial blood.

25

What is unique about the protein buffer system? _____

_____.

This system carries a(n) _____ and _____ charge.

— — — — — — — — — — — — —

It is amphoteric—it has the ability to bind or release excess H$^+$ ions.
acidic and basic

26

The four buffer systems in the body are:

1. _____

2. _____

3. _____

4. _____

_ _ _ _ _ _ _ _ _ _ _ _ _ _ _

1. Bicarbonate–carbonic acid
2 Phosphate
3. Hemoglobin–oxyhemoglobin
4. Protein

Table 15B gives explanations of the Ion Exchange and the Respiratory Regulation as regulatory mechanisms for pH control. The ion exchange is frequently referred to as the Chloride Shift. Refer to Chapter 2, frame 166 for clarification of the chloride shift if needed. The respiratory regulation depends upon the lungs in exhaling CO_2 or retaining the CO_2 for the control of pH. Again the interventions refer to the action and rationale refers to the reason.

Study this table carefully and refer back to it as needed.

TABLE 15   The Regulatory Mechanism for pH Control.
B. Ion Exchange and Respiratory Regulation

Regulatory Mechanism	Intervention	Rationale
Ion exchange	The ion exchange of HCO_3 and Cl occurs in the red blood cell as the result of the O_2 and CO_2 exchange. There is a redistribution of anions in response to an increase in the CO_2. The chloride ion enters the red blood cell as the bicarbonate ion diffuses into the plasma in order to restore ionic balance.	An increase or carbon dioxide serum concentration causes the CO_2 to diffuse into the red blood cells combining with H_2O to form H_2CO_3. This weak acid dissociates forming acid and base ions, $H_2CO_3 \rightarrow H^+ + HCO_3^-$. The hydrogen ion is buffered by the hemoglobin and the HCO_3^- ion moves into the plasma as the chloride ion (Cl^-) shifts into the cell to replace it.
Respiratory regulation (acts quickly in case of an emergency)	For regulation of acid, the lungs will blow off more CO_2 and for regulation of alkaline, the respiratory center will be depressed in order to retain the CO_2.	The respiratory center in the medulla controls the rate and depth of respiration and is sensitive to changes in blood pH or CO_2 concentration. When the pH is decreased, carbonic acid is exhaled in the form of carbon dioxide and moist air ($H_2CO_3 \rightarrow H_2O + CO_2$), thus hydrogen ions are also eliminated.

27

When carbon dioxide enters the red blood cell, what happens to the bicarbonate and chloride anions? _____

_____ .

  This ion exchange is frequently called the *_____ .

_ _ _ _ _ _ _ _ _ _ _ _ _ _ _

Bicarbonate diffuses out of the cell and the chloride ion enters the cell.
chloride shift

28

When the red blood cells are oxygenated, what anion is commonly

present? _____ .

  When carbon dioxide enters the red blood cells, what anion also

enters? _____ .

  What anion leaves as carbon dioxide enters? _____ .

_ _ _ _ _ _ _ _ _ _ _ _ _ _ _

bicarbonate (HCO_3), chloride (Cl), bicarbonate (HCO_3)

29

Which regulatory mechanism acts most quickly in an emergency situation? _____ .

_ _ _ _ _ _ _ _ _ _ _ _ _ _ _

respiratory regulatory

30

How does the respiratory regulatory mechanism control the serum pH?

_____

_____ .

– – – – – – – – – – – – – – –

When the serum pH is decreased, the lungs will blow off CO_2. With an increased pH, the lungs retain CO_2.

31

Where is the respiratory center located? _____ . What does

the respiratory center control? _____ .

– – – – – – – – – – – – – –

medulla, rate and depth of the respiration

32

How do the two mechanisms in Table 15B deal with an increased CO_2?

_____

_____ .

– – – – – – – – – – – – – – –

1. As the CO_2 enters the red cells, HCO_3 leaves the red cell and Cl enters.
2. The lungs will blow off more CO_2.

Table 15C describes how the kidneys regulate pH in the body. The kidneys compensate for an excess production of acid by excreting the acid and returning the bicarbonate to the extracellular fluid. The acid occurs as the result of normal metabolism.

Study the table carefully, noting how in each instance the excess hydrogen ions are neutralized. When you think you understand the renal regulatory mechanism described in the table, answer the frames which follow. Refer back to the table when needed.

TABLE 15 The Regulatory Mechanism for pH Control. C. Renal Regulation

Regulatory Mechanism	Intervention	Rationale
a. Acidification of phosphate buffer salts	The exchange mechanism occurs between the H^+ of the renal tubular cells and the sodium salt (Na_2HPO_4) in the tubular urine.	The sodium salt (Na_2HPO_4) dissociates into Na^+ and $NaHPO_4^-$ ions; sodium ion moves into the tubular cell and the hydrogen unites with the $NaHPO_4$ forming a dehydrogen phosphate salt, NaH_2PO_4, which is excreted.
b. Reabsorption of bicarbonate	Carbon dioxide is absorbed by the tubular cells from the blood and combines with water present in the cells to form carbonic acid which in turn ionizes forming H^+ and HCO_3^-. The Na^+ ion of the tubular urine exchanges with the H^+ ion of the tubular cells and combines with HCO_3^- to form sodium bicarbonate $NAHCO_3$, which is reabsorbed into the blood.	The enzyme, carbonic anhydrase, is responsible for the formation of carbonic acid H_2CO_3. The ionization of H_2CO_3 $\rightarrow H^+ + HCO_3^-$. The free hydrogen ion will exchange with the sodium ion. The exchange of the H^+ ion and the Na^+ ion permits the reabsorption of a bicarbonate with sodium ($NaHCO_3$) and the excretion of an acid or the H^+ ion.
c. Secretion of ammonia	Ammonia (NH_3) unites with HCl in the renal tubules and the H^+ ion is excreted as NH_4Cl (ammonium chloride).	Almost half of the H^+ ion excretion is from this method—$HCl + NH_3$ $\rightarrow NH_4Cl$. Ammonia is formed in the renal tubular cells by oxidative breakdown of the amino acid glutamine in the presence of the enzyme glutaminase. Ammonia can also be converted into urea by the liver and excreted as urea by the kidneys.

33

Name the three renal regulatory mechanisms for pH control.

1.

2.

3.

– – – – – – – – – – – – – – –

1. Acidification of phosphate buffer salts
2. Reabsorption of bicarbonate
3. Secretion of ammonia

34

The hydrogen ions exist largely in the renal tubules in the buffered state. They are excreted indirectly, replacing a cation in excretion.

   Explain how the hydrogen ion is excreted after it combines with the

sodium salt—Na_2HPO_4. _____

_____.

– – – – – – – – – – – – – –

The H^+ ion replaces a Na^+ forming a dehydrogen phosphate salt— NaH_2PO_4. This salt is excreted.

35

Explain how the hydrogen ion is derived from carbonic acid. _____

_____.

   Explain how the bicarbonate ion is reabsorbed. _____

_____.

– – – – – – – – – – – – – –

The ionization of $H_2CO_3 \rightarrow H + HCO_3$
The Na^+ ion exchanges with the H^+ ion in renal tubules and the Na^+ ion combines with HCO_3 and is reabsorbed into the blood.

36

Explain the formation of ammonia in the renal tubular cells. _____

_____ .

How does the ammonia aid in the excretion of the hydrogen ion?

_____ .

– – – – – – – – – – – – – – –

Formed from the breakdown of amino acid glutamine.

NH_3 unites with HCl and is excreted as NH_4Cl (ammonium chloride).

37

Which one of the three methods in Table 15C is responsible for nearly

half of the hydrogen ion excretion? _____ .

– – – – – – – – – – – – – – –

secretion of ammonia

38

The four regulatory mechanisms for pH control are:

1. _____

2. _____

3. _____

4. _____

The principal buffer system of the body is *_____ .

– – – – – – – – – – – – – – –

1. Buffer system
2. Ion exchange
3. Respiratory regulation
4. Renal regulation
bicarbonate–carbonic acid

Review

Q. 1

The acidity or alkalinity of a solution depends upon the concentration of the * _____.

_ _ _ _ _ _ _ _ _ _ _ _ _ _ _

hydrogen ion

Q. 2

The pH of extracellular fluid in health is between _____.

A serum pH below 7.35 would indicate _____.

_ _ _ _ _ _ _ _ _ _ _ _ _ _ _

7.35 and 7.45, acidosis

Q. 3

Define the following terms:
1. Anion.
2. Cation
3. Acid.
4. Base

_ _ _ _ _ _ _ _ _ _ _ _ _ _ _

1. negative charged ion
2. positive charged ion
3. donates a proton, or serum pH below 7.35 (excess hydrogen)
4. accepts a proton, or serum pH above 7.45 (excess bicarbonate)

Q. 4

As the result of buffering, the strong acid is replaced by a weak acid and a neutral salt, therefore, fewer _____ ions will be released.

_ _ _ _ _ _ _ _ _ _ _ _ _ _ _

hydrogen

Q. 5
Explain how the phosphate buffer system removes the excess hydrogen

ions from the body. _____

_____ .

The sodium ion, from exchange, combines with the bicarbonate ion
and is reabsorbed into the extracellular fluid; thus, the extracellular
fluid would become more (acid/alkaline).

_ _ _ _ _ _ _ _ _ _ _ _ _ _ _

H^+ exchanges with the Na^+ ion of Na_2HPO_4 and is excreted as
NaH_2PO_4, alkaline

Q. 6
What buffer system maintains the same pH level in the venous blood as

in the arterial blood? _____ .

_ _ _ _ _ _ _ _ _ _ _ _ _ _ _

hemoglobin–oxyhemoglobin

Q. 7
Proteins are amphoteric. Explain what they are able to do with the

excess hydrogen ion. _____

_____ .

_ _ _ _ _ _ _ _ _ _ _ _ _ _ _

Proteins are able to bind or release excess H^+ ions.

Q. 8
With ion exchange, there is a redistribution of anions. Name these two

anions. _____ and _____ . When does this shift

occur? _____ .

_ _ _ _ _ _ _ _ _ _ _ _ _ _

chloride and bicarbonate, When CO_2 enters red blood cell.

Q. 9
Why do the lungs blow off more CO_2? _____.

_ _ _ _ _ _ _ _ _ _ _ _ _ _ _ _ _

to eliminate the H^+ ion

Q. 10
Ammonia unites with hydrogen chloride and is excreted as *_____

_____. The pH of the urine would be (acid/ alkaline).

_ _ _ _ _ _ _ _ _ _ _ _ _ _ _ _

NH_4Cl–ammonium chloride, acid

Diagram 8 demonstrates the normal acid-base balance or equilibrium and the cause and results of imbalance. The top scale shows acid-base balance with 1 part acid and 20 parts base. To determine the acid-base imbalance, two variables must be known: the respiratory component, or plasma CO_2, and the metabolic component, or plasma HCO_2.

Respiratory acidosis has an increased plasma CO_2 because of a decrease in alveolar ventilation. The CO_2 combines with water forming carbonic acid, H_2CO_3.

Respiratory alkalosis has a decreased plasma CO_2, because of an increase in alveolar ventilation.

Metabolic acidosis has a decreased plasma CO_2 because of the gain of acid or loss of bicarbonate from the extracellular fluid. The plasma bicarbonate is expressed as plasma CO_2 or CO_2 combining power which is the end product of metabolism.

Metabolic alkalosis has an increased plasma CO_2 because of the loss of a strong acid from the body.

In many cases it is necessary to test the pH of the blood to fully evaluate the acid-base imbalance.

Study the diagram carefully and refer back to the diagram or this introduction as needed.

Plasma CO_2 is a serum bicarbonate determinant and is frequently called CO_2 combining power. It refers to the amount of cations, e.g.,

H^+, Na^+, K^+, etc., available to combine with HCO_3^-. The level of HCO_3 in the blood is determined by the amount of CO_2 dissolved in the blood. Refer to Diagram 7 if necessary for further clarification.

Left Side
Related to Respiratory Function

Right Side
Related to Metabolism of the Body

Balance

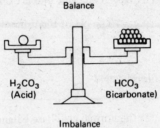

H_2CO_3
(Acid)

HCO_3
Bicarbonate)

Imbalance

Respiratory Acidosis

Metabolic Acidosis

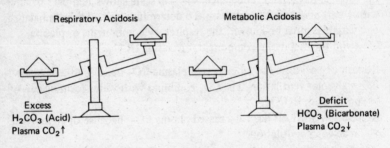

Excess
H_2CO_3 (Acid)
Plasma $CO_2\uparrow$

Deficit
HCO_3 (Bicarbonate)
Plasma $CO_2\downarrow$

Respiratory Alkalosis

Metabolic Alkalosis

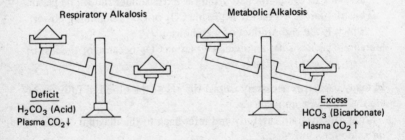

Deficit
H_2CO_3 (Acid)
Plasma $CO_2\downarrow$

Excess
HCO_3 (Bicarbonate)
Plasma $CO_2\uparrow$

Diagram 8. Acid-Base Balance and Imbalance

39

In acid-base balance the left side or the carbonic acid side is controlled by:
( ) a. respiratory function
( ) b. body metabolism
The right side or the base bicarbonate is controlled by:
( ) a. respiratory function
( ) b. body metabolism

— — — — — — — — — — — — — —

a. X   a. —
b. —   b. X

40

With respiratory acidosis, there is *_____

_____ .

With respiratory alkalosis, there is *_____

_____ .

With metabolic acidosis, there is *_____

_____ .

With metabolic alkalosis, there is *_____

_____ .

— — — — — — — — — — — — — —

carbonic acid excess (also increased plasma CO_2)
carbonic acid deficit (also decreased plasma CO_2)
base bicarbonate deficit (also decreased plasma CO_2)
base bicarbonate excess (also increased plasma CO_2)

41

Respiratory acidosis and metabolic alkalosis will cause a(n) (increase/ decrease) in serum CO_2.
Respiratory alkalosis and metabolic acidosis will cause a(n) (increase/ decrease) in serum CO_2.

_ _ _ _ _ _ _ _ _ _ _ _ _ _ _

increase, decrease

42

Place CE—carbonic acid excess
Place CD—carbonic acid deficit
Place BE—base bicarbonate excess
Place BD—base bicarbonate deficit
concerning the following conditions:

_____ a. Metabolic acidosis

_____ b. Metabolic alkalosis

_____ c. Respiratory acidosis

_____ d. Respiratory alkalosis

_____ , _____ e. Serum (plasma) CO_2 increased

_____ , _____ f. Serum (plasma) CO_2 decreased

_ _ _ _ _ _ _ _ _ _ _ _ _ _ _

a. BD
b. BE
c. CE
d. CD
e. BE, CE
f. BD, CD

Clinical Considerations

Table 16 gives various clinical conditions that can cause metabolic acid-osis and metabolic alkalosis. Study this table carefully, until you under-stand and can state the reasons these conditions can cause acidosis or alkalosis, then proceed to the frames. Refer back to this table as needed.

TABLE 16   Conditions Causing Metabolic Acidosis and Metabolic Alkalosis

Conditions	Metabolic Acidosis	Metabolic Alkalosis
Uncontrolled diabetes mellitus (diabetic acid-osis)	With the failure to metabolize adequate quantities of glucose, the liver will increase the metabo-lism of fatty acids. The oxidation of fatty acids produces ketone bodies. H^+ ions circulate with the ketones which make the blood more acid. Every ketone requires a base for excretion.	
Severe diarrhea	The loss of sodium ions exceeds that of chloride ions. The kidney mechanisms for conserving sodium and water and for ex-creting the hydrogen ion fail.	
Starvation, kidney failure, excessive exer-cise, severe infections	Acid metabolites occur as the result of an abnormal accumu-lation of acid products from metabolism or cellular break-down (from starvation). These acids neutralize alkali, and the excess acid then causes the balance to tilt left.	
Stomach ulcer (peptic ulcer)		An excess of alkali in the extracellular fluid occurs when the patient takes ex-cessive amounts of baking soda, $NaHCO_3$, to ease his ulcer pain.

TABLE 16 (continued)

Conditions	Metabolic Acidosis	Metabolic Alkalosis
Vomiting, gastric suction, mercurial diuretics		Large quantities of chloride are lost. Bicarbonate anions increase to compensate for the chloride loss. The number of anions must be equal to the number of cations.
Loss of potassium		Loss of potassium from the body is accompanied by loss of chloride. (As stated in Chapter 2 on chlorides.)

43

With uncontrolled diabetes mellitus, glucose cannot be metabolized;

therefore, what will occur? _____

_____

_ _ _ _ _ _ _ _ _ _ _ _ _ _ _

The liver will produce fatty acids and later ketone bodies will be produced.

44

What makes the blood more acid in diabetic acidosis? _____

_____ .

Why is there a base deficit? _____

_____ .

_ _ _ _ _ _ _ _ _ _ _ _ _ _

H^+ ions circulating with the ketone bodies which are also acid.
The base bicarbonate is excreted with the ketone bodies.

45

In severe diarrhea, explain which ion is excreted in excess? _____.

Explain how this can cause an acidotic state. _____

_____ .

_ _ _ _ _ _ _ _ _ _ _ _ _ _

sodium ion
The Cl^- ion is retained and combines with the H^+ ion, producing HCl.

46

How can starvation contribute to metabolic acidosis? _____

_____ .

_ _ _ _ _ _ _ _ _ _ _ _ _ _

Acid metabolites resulting from cellular breakdown.

47

How can excessive exercise, kidney failure, and severe infections con-

tribute to metabolic acidosis? _____

_____ .

_ _ _ _ _ _ _ _ _ _ _ _ _ _

An abnormal accumulation of acid products from metabolism

48
Indicate which of the following conditions can cause metabolic acid-osis—a deficit of alkali or base bicarbonate.

( ) a. Starvation
( ) b. Gastric suction
( ) c. Excessive exercise
( ) d. Mercurial diuretics
( ) e. Severe infections
( ) f. Uncontrolled diabetes mellitus

– – – – – – – – – – – – – – – –

a. X
b. —
c. X
d. —
e. X
f. X

49
When a patient takes excessive amounts of baking soda to ease indigestion

or stomach ulcer pain, what will probably occur? _____ .

Why? _____ .

– – – – – – – – – – – – – – – –

metabolic alkalosis,
There is an excess alkali in the extracellular fluid

50
Name the anion that is lost in great quantities due to vomiting, gastric

suction, or mercurial diuretics. _____ .

– – – – – – – – – – – – – – – –

chloride

51

Name the conditions that cause metabolic alkalosis.

1.
2.
3.
4.
5.

_ _ _ _ _ _ _ _ _ _ _ _ _ _ _

1. treated stomach ulcer
2. vomiting
3. gastric suction
4. mercurial diuretics
5. loss of potassium

52

Place BE for base bicarbonate excess or metabolic alkalosis and BD for base bicarbonate deficit or metabolic acidosis:

_____ a. Uncontrolled diabetes mellitus

_____ b. Treated stomach ulcer

_____ c. Severe diarrhea

_____ d. Severe infection

_____ e. Vomiting, gastric suction

_____ f. Mercurial diuretics

_____ g. Fever, severe infections

      h. Excessive exercise

_ _ _ _ _ _ _ _ _ _ _ _ _ _ _

a. BD	e. BE
b. BE	f. BE
c. BD	g. BD
d. BD	h. BD

Table 17 gives the clinical conditions and causes of respiratory acidosis and respiratory alkalosis. Study this table carefully and be prepared to explain why respiratory acidosis or alkalosis would occur with these conditions or situations. Proceed to the frames that follow and refer to the table whenever necessary.

TABLE 17   Conditions Causing Respiratory Acidosis and Respiratory Alkalosis

Conditions	Respiratory Acidosis	Respiratory Alkalosis

Central Nervous System Depressants
 1. Narcotics*
   a. Morphine
   b. Meperidine hydrochloride (trade name is Demerol)
 2. Anesthetics*
 3. Barbiturates* (in large doses)
	These drugs depress the respiratory center in the medulla causing a retention of CO_2 (carbon dioxide) in the blood.
	
	 These produce an increase in carbonic acid in the blood plasma.	

Pulmonary Dysfunctions 1. Emphysema 2. Bronchiectasis 3. Pneumonia 4. Poliomyelitis	An inadequate exchange of gases in the lungs due to a decrease in aeration surface causes a retention of CO_2 in the blood.	
Hyperventilation 1. Anxiety-Hysteria 2. Drug toxicity, aspirin 3. Diseases, tetany 4. Fever 5. Swimming 6. Deliberate overbreathing		An excessive blowing off of CO_2 through the lungs reduces the amount of carbonic acid in the extracellular fluid

*Existing when taken in large quantities and/or over an extended time.

53

Explain how an inadequate exchange of gases in the lungs can cause

respiratory acidosis? _____

_____ .

_ _ _ _ _ _ _ _ _ _ _ _ _ _ _ _

It can cause a retention of CO_2 in the blood. H_2O and $CO_2 \rightarrow H_2CO_3$

54

Which of the following will depress the respiratory center causing carbon
dioxide retention?

( ) a. Morphine
( ) b. Demerol
( ) c. Hyperventilation
( ) d. Anesthesia
( ) e. Barbiturates
( ) f. Aspirin

_ _ _ _ _ _ _ _ _ _ _ _ _ _ _ _

a. X
b. X
c. —
d. X
e. X
f. —

55

What four conditions might lead to respiratory acidosis due to a decrease
in aeration surface in the lung?

1.
2.
3.
4.

_ _ _ _ _ _ _ _ _ _ _ _ _ _ _

1. emphysema   3. bronchiectasis
2. pneumonia   4. poliomyelitis

56

What is respiratory alkalosis? _____

_____ .

— — — — — — — — — — — — — —

decreased plasma CO_2 due to increased ventilation—excessive blowing off CO_2

57

Hyperventilation can result from:
1.
2.
3.
4.
5.
6.

— — — — — — — — — — — — — —

1. anxiety-hysteria
2. drug toxicity, aspirin
3. diseases, tetany
4. fever
5. swimming
6. deliberate overbreathing

58

Explain the difference between respiratory alkalosis and respiratory

acidosis. _____

_____ .

— — — — — — — — — — — — — —

Respiratory alkalosis—decreased plasma CO_2. Respiratory acidosis—increased plasma CO_2 and/or other descriptions.

59

List the conditions or situations which may cause respiratory acidosis and respiratory alkalosis.

Respiratory Acidosis	Respiratory Alkalosis
1.	1.
2.	2.
3.	3.
4.	4.
5.	5.
6.	6.
7.	

— — — — — — — — — — — — — —

Respiratory Acidosis	Respiratory Alkalosis
1. narcotics	1. anxiety
2. anesthetics	2. drug toxicity
3. barbiturates	3. tetany
4. emphysema	4. fever
5. bronchiectasis	5. swimming
6. pneumonia	6. overbreathing
7. poliomyelitis	

Review

Q. 1

What will happen to plasma CO_2 if there is an excess of carbonic acid and an excess of base bicarbonate? _____ .

— — — — — — — — — — — — — —

increased plasma CO_2

Q. 2

What will happen to plasma CO_2 is there is a deficit of carbonic acid and a deficit of base bicarbonate? _____ .

— — — — — — — — — — — — — —

decreased plasma CO_2

Q. 3

How does uncontrolled diabetes mellitus cause metabolic acidosis?

_____

_____ .

_ _ _ _ _ _ _ _ _ _ _ _ _ _ _

Failure to metabolize glucose produces ketone bodies, which are acid, and an increase of H^+ ions.

Q. 4

In severe cases of diarrhea, the sodium ion loss is greater than the loss

of _____ ion.

_ _ _ _ _ _ _ _ _ _ _ _ _ _ _

chloride

Q. 5

Gastric suctioning can cause a base bicarbonate excess. Why? _____

_____ .

_ _ _ _ _ _ _ _ _ _ _ _ _ _ _

Bicarbonate ions increase to compensate for the chloride loss.

Q. 6

Taking too much sodium bicarbonate for treatment of indigestion can

cause *_____ .

_ _ _ _ _ _ _ _ _ _ _ _ _ _ _

metabolic alkalosis

Q. 7

Drugs that depress the respiratory center will cause a(n) (excretion/ retention) of CO_2.

_ _ _ _ _ _ _ _ _ _ _ _ _ _ _

retention

Q. 8

With inadequate exchange of gases in the lungs, respiratory (acidosis/ alkalosis) can occur.

— — — — — — — — — — — — — —

acidosis

Q. 9

Acute anxiety can cause hyperventilation, resulting in *_____

_____.

— — — — — — — — — — — — — —

respiratory alkalosis or carbonic acid deficit

Table 18 lists selected symptoms of metabolic acidosis and alkalosis, and respiratory acidosis and alkalosis. Study these symptoms carefully. Memorize the plasma CO_2 and pH changes for each of the imbalances. Refer to the glossary for any unknown words. Refer to this table as you find it necessary.

You may find it necessary to refer to Diagram 7 for clarification of pH changes, Diagram 8 for clarification of plasma CO_2, and Table 15 for the regulatory mechanisms for compensation of acid-base imbalance.

TABLE 18   Selected Symptoms of Metabolic Acidosis and Alkalosis,
Respiratory Acidosis and Alkalosis

Metabolic Acidosis	Metabolic Alkalosis
Kussmaul breathing–hyperactive, abnormally vigorous breathing.	Shallow breathing.
Flushing of the skin–capillary dilatation due to CO_2 accumulation in the tissues.	Vomiting–loss of Cl^- ion and loss of K^+ ion.
Dehydration–water loss via kidney and later through vomiting.	pH increased–when compensatory mechanisms fail.
Restlessness	Plasma CO_2 ↑
pH decreased–when compensatory mechanisms fail.	
Plasma CO_2 ↓	
Death–from renal and respiratory failure	

Respiratory Acidosis	Respiratory Alkalosis
Dyspnea–labored or difficult breathing.	Overbreathing–rapid and shallow due to hydrogen ion or CO_2 deficit.
Disorientation	Vertigo (dizziness)–overbreathing.
pH is decreased–when compensatory mechanisms fail.	Tetany spasms
Plasma CO_2 ↑	Unconsciousness–later
	pH is increased–when compensatory mechanisms fail.
	Plasma $CO 2$ ↓

60

Metabolic acidosis results from a *_____.

What happens to the plasma CO_2? _____.

– – – – – – – – – – – – – – –

base bicarbonate deficit. It decreases.

61

The common symptoms of metabolic acidosis are:

1.

2.

3.

4.

— — — — — — — — — — — — — — — —

1. hyperactive, vigorous breathing
2. dehydration
3. flushing of the skin
4. restlessness

62

With metabolic acidosis, the renal and respiratory mechanisms try to re-establish balance.

Explain how the renal mechanism works to reestablish balance.

_____.

Explain how the respiratory mechanism works to reestablish balance.

_____.

When these two mechanisms fail, what happens to the plasma pH?

_____.

— — — — — — — — — — — — — — —

The H^+ ion exchanges for the Na^+ ion and thus H^+ ion is excreted.
As the result of the Kussmaul breathing CO_2 is blown off. $H_2O + CO_2$
$\rightarrow H_2CO_3$.
It decreases.

63

Metabolic alkalosis results from a *_____

_____. What happens to the plasma CO_2? _____.

— — — — — — — — — — — — — —

base bicarbonate excess, It increases.

64

The common symptoms of metabolic alkalosis are:
1.
2.

— — — — — — — — — — — — — — —

1. shallow breathing
2. vomiting

65

With metabolic alkalosis, the buffer, renal, and respiratory mechanisms will try to reestablish balance. With the buffer mechanism, the excess bicarbonate reacts with buffer acid salts, thus, there will be a decrease of bicarbonate in the extracellular fluid and an increase of carbonic acid.

Explain how the renal mechanism works to reestablish balance.

_____ .

Explain how the respiratory mechanism works to reestablish balance.

_____ .

When these three mechanisms fail, what happens to the plasma pH?

_____ .

— — — — — — — — — — — — — — —

H^+ ion is conserved and Na^+ and K^+ ions are excreted with HCO_3.
Pulmonary ventilation is decreased, therefore, CO_2 is retained, producing H_2CO_3.
It increases.

66

Respiratory acidosis results from a *_____ . What

happens to the plasma CO_2? _____ .

— — — — — — — — — — — — — — —

carbonic acid excess, It increases.

67

The common symptoms of respiratory acidosis are:
1.
2.

— — — — — — — — — — — — — —

1. dyspnea
2. disorientation

68

With respiratory acidosis, the buffer, renal, and respiratory mechanisms will try to reestablish balance. As the result of the chloride shift, bicarbonate ions are released to neutralize carbonic acid excess.

With an increased plasma CO_2, explain how the respiratory mechanism works to compensate for this imbalance. _____

_____.

Explain how the renal mechanism works to compensate for this imbalance. _____

_____.

When these three mechanisms fail, what happens to the plasma pH?

_____.

— — — — — — — — — — — — — — —

CO_2 will stimulate the respiratory center to increase the rate and depth of respiration. CO_2 is blown off with water. $H_2O + CO_2 \rightarrow H_2CO_3$.
1. The H^+ ion exchanges with the Na^+ ion and the Na^+ is reabsorbed with the HCO_3.
2. An increased formation of ammonia.
It is decreased.

69
Respiratory alkalosis results from a * _____ .

What happens to the plasma CO_2? _____ .

– – – – – – – – – – – – – – –

carbonic acid deficit. It decreases.

70
The common symptoms of respiratory alkalosis are:
1.
2.
3.
4.

– – – – – – – – – – – – – – –

1. rapid and shallow breathing
2. vertigo
3. tetany spasms
4. unconsciousness

71
Frequently, with respiratory alkalosis, you will notice that these individuals are very apprehensive and anxious. They will hyperventilate to overcome their anxiety. Many times this occurs from a psychological reason, e.g., giving a speech for the first time, fear of failing an exam. How do you think you might help the respiratory mechanism compensate for this imbalance? _____

_____ .

– – – – – – – – – – – – – – –

There is a lack of plasma CO_2, so giving CO_2, e.g., rebreathing CO_2 from a paper bag, would stimulate the lungs to breathe deeper and then slower.

72

The buffer mechanism produces more organic acids, in respiratory alkalosis, which react with the excess bicarbonate ion.

How do you think the renal mechanism works to compensate for this

imbalance? _____.

When these three mechanisms fail, what happens to the plasma pH?

_____.

– – – – – – – – – – – – – – –

An increased HCO_3 excretion and a H^+ ion retention. It increases

73

Place M. Ac for metabolic acidosis,
Place M. Al for metabolic alkalosis,
Place R. Ac for respiratory acidosis,
Place R. Al for respiratory alkalosis,
concerning the following symptoms:

_____ a. Hyperactive, vigorous breathing or Kussmaul breathing

_____ b. Dyspnea

_____ c. Shallow breathing

_____ d. Rapid, shallow breathing (overbreathing)

_____ e. Flushing of the skin

_____ f. Dehydration

_____ g. Vomiting

_____ h. Vertigo

_____ i. Tetany spasms

_____ j. Disorientation

a. M. Ac	f. M. Ac
b. R. Ac	g. M. Al
c. M. Al	h. R. Al
d. R. Al	i. R. Al
e. M. Ac	j. R. Ac

Diagram 9 gives the four acid-base imbalances, explains the body's defense action against these imbalances, and gives various methods of
treatment for restoring balance. The frames following Table 18 had you
think of the various regulatory mechanisms from Table 15 A, B, and C,
used for restoring acid-base balance. This diagram describes briefly the
respiratory and renal mechanisms for restoring balance in respect to the
treatment given. The diagram should aid in reinforcing your understanding of these two mechanisms used for our body's defense action.

Study this diagram carefully, with particular attention to the cause of each imbalance, the body's defense action, the pH of the urine as to whether it is acid or alkaline, and the treatment for these imbalances. Refer to these diagrams and tables whenever you find it necessary.

Metabolic Acidosis (Deficit of bicarbonate in the extracellular fluid)

Lungs   Kidney

Lungs "blow off" acid. Respirations are increased.

Urine is acid. Kidneys conserve alkali and excrete acid.

Treatment: Remove the cause. Administer an I.V. alkali solution, e.g., sodium bicarbonate or sodium lactate. Restore water, electrolytes, and nutrients.

Metabolic Alkalosis (Excess of bicarbonate in the extracellular fluid)

Lungs Kidney

Breathing is suppressed.

Urine is alkaline. Kidneys excrete alkali ions, and retain hydrogen ions and nonbicarbonate ions.

Treatment: Remove the cause. Administer an I.V. solution of chloride, e.g., sodium chloride. Replace potassium deficit.

Respiratory Acidosis (Excess of carbonic acid in the extracellular fluid)

Kidney

Lungs are affected. However, H_2CO_3 will stimulate the lungs to "blow off" acid, by breathing deeper.

Urine is acid. Kidneys conserve alkali; excrete acid.

Treatment: Remove the cause. Administer an I.V. alkali solution. Deep breathing exercise.

Respiratory Alkalosis (Deficit of carbonic acid in the extracellular fluid)

Kidney

Lungs are affected. Treatment would be recommended.

Urine is alkaline. Kidneys excrete alkali; retain acid.

Treatment: Remove the cause. Administer an I.V. solution of chloride. The chloride will replace the alkali. Rebreathe expired air, e.g., CO_2, from a paper bag.

Diagram 9. Body's Defense Action and Treatment for Acid-Base Imbalance

74

What is metabolic acidosis? _____ .

What happens to the urine? _____ .

What are the body's defense actions against it?
1.
2.

— — — — — — — — — — — — — —

base bicarbonate deficit, acid
1. lungs blow off CO_2 or acid
2. kidneys excrete acid or H^+ ion and conserve alkali

75

How is metabolic acidosis treated? _____

_____

_____ .

— — — — — — — — — — — — — —

1. remove cause
2. administer I.V. alkali solution, e.g., $NaHCO_3$
3. restore H_2O and electrolyte

76

What is metabolic alkalosis? _____ .

What happens to the urine? _____ .

What are the body's defense actions against it?
1.
2.

— — — — — — — — — — — — — —

base bicarbonate excess, alkaline
1. breathing is suppressed
2. kidneys excrete alkali ions, e.g., HCO_3 and retain H^+ ions and nonbicarbonate ions

77

How is metabolic alkalosis treated? _____ .

_____

_____ .

– – – – – – – – – – – – – –

1. remove cause
2. administer I.V. chloride solution, e.g., NaCl
3. replace K deficit

78

What is respiratory acidosis? _____ .

What happens to the urine? _____ .

What are the body's defense action against it?
1.
2.

– – – – – – – – – – – – – –

carbonic acid excess, acid
1. Excess CO_2 stimulates the lung to blow off CO_2 or acid
2. Kidneys conserve alkali and excrete H^+ ions or acid urine

79

How is respiratory acidosis treated? _____

_____

_____ .

– – – – – – – – – – – – – –

1. remove cause
2. administer I.V. alkali solution, e.g., $NaHCO_3$
3. deep breathing exercise

80

What is respiratory alkalosis? _____ .

  What happens to the urine? _____ .

What is the body's defense action against it? _____

_____ .

‒ ‒ ‒ ‒ ‒ ‒ ‒ ‒ ‒ ‒ ‒ ‒ ‒ ‒

carbonic acid deficit, alkaline, Kidneys excrete alkaline HCO_3, and retain acid, H^+

81

How is respiratory alkalosis treated? _____

_____

_____ .

‒ ‒ ‒ ‒ ‒ ‒ ‒ ‒ ‒ ‒ ‒ ‒ ‒ ‒

1. remove cause
2. administer an I.V. chloride solution, NaCl
3. rebreathe expired air

Review

Q. 1

Hyperactive, vigorous breathing, called Kussmaul breathing, is a symptom of *_____ .

‒ ‒ ‒ ‒ ‒ ‒ ‒ ‒ ‒ ‒ ‒ ‒ ‒ ‒

metabolic acidosis

Q. 2

Dyspnea, decreased pH, and increased serum CO_2 are symptoms of

*_____ .

‒ ‒ ‒ ‒ ‒ ‒ ‒ ‒ ‒ ‒ ‒ ‒ ‒ ‒

respiratory acidosis

Q. 3

In metabolic acidosis, the lungs will *_____ and the kidneys will excrete the *_____ .

— — — — — — — — — — — — —

"blow off" CO_2, H^+ ions or acid

Q. 4

In metabolic alkalosis, the lungs will *_____ and the kidneys will excrete *_____ .

— — — — — — — — — — — — —

retain CO_2, the alkali ions

Q. 5

In respiratory acidosis, the kidneys will conserve the _____ ions and excrete _____ .

— — — — — — — — — — — — — —

alkali or HCO_3, acid or H^+ ions

Q. 6

In respiratory alkalosis, the kidneys will conserve the _____ ions and excrete _____ .

— — — — — — — — — — — — —

H^+ or acid, alkali or HCO_3

Q. 7

Administration of an I.V. alkali solution is the treatment for *_____ _____and _____ .

— — — — — — — — — — — — —

respiratory acidosis, metabolic acidosis

Q. 8
Give an example of an alkali solution. _____.

$- - - - - - - - - - - - - - - -$

NaHCO$_3$ or sodium bicarbonate solution

Q. 9
Administration of an I.V. chloride solution is the treatment for

*_____ and _____.

$- - - - - - - - - - - - - - - -$

respiratory alkalosis, metabolic alkalosis

Q. 10
Give an example of a chloride solution _____.

$- - - - - - - - - - - - - - - -$

NaCl or sodium chloride solution

Q. 11

Complete the following chart on acid-base imbalance as to the plasma CO_2 findings and the plasma pH findings. Use the arrow pointed upward for increase and the arrow pointed downward for decrease.

Metabolic Acidosis	Metabolic Alkalosis
Plasma CO_2	Plasma CO_2
Plasma pH	Plasma pH

Respiratory Acidosis	Respiraltory Alkalosis
Plasma CO_2	Plasma CO_2
Plasma pH	Plasma pH

Metabolic Acidosis	Metabolic Alkalosis
Plasma CO_2 ↓	Plasma CO_2 ↑
Plasma pH ↓	Plasma pH ↑

Respiratory Acidosis	Respiratory Alkalosis
Plasma CO_2 ↑	Plasma CO_2 ↓
Plasma pH ↓	Plasma pH ↑

SUMMARY QUESTIONS

The questions in this summary will test your comprehensive ability and knowledge of the material found in Chapter 3. If the answer is unknown, refer to the section in the chapter for further understanding and clarification.

1. The acidity of a solution is determined by the concentration of the

_____.

2. Define the symbol pH and OH^-. What is the pH of extracellular fluid

in health? Neutral solutions have a pH of _____.

3. What are the four regulatory mechanisms for pH control and explain briefly how each functions.

4. Define metabolic acidosis and alkalosis, and respiratory acidosis and alkalosis.

5. Name several conditions which can cause metabolic acidosis and alkalosis, and respiratory acidosis and alkalosis.
6. Name at least two symptoms of metabolic acidosis and alkalosis, and respiratory acidosis and alkalosis.
7. Explain the body defense action and treatment for the four acid-base imbalances.

References

Abbott Laboratories, *Fluid and Electrolytes*, North Chicago, Ill., 1968, pp. 22–24.

Barbata, Jean C., *et al.*, *Medical-Surgical Nursing*, New York: G. P. Putnam's Sons, 1964, pp. 519–520.

Baxter Laboratories, Inc., *Fluid Therapy*, Morton Grove, Ill., 1962, pp. 59–67.

Burgess, Richard, "Fluids and Electrolytes," *American Journal of Nursing*, LXV (October, 1965), p. 92.

Dutcher, Isabel E. and Sandra B. Fielo, *Water and Electrolytes*, New York: The Macmillan Co., 1967, pp. 30–39, 52–57.

Guyton, Arthur C., *Function of the Human Body*, 2nd ed., Philadelphia, Pa.: W. B. Saunders Co., 1964, pp. 78–82.

Guyton, Arthur C., *Textbook of Medical Physiology*, 3rd ed., Philadelphia, Pa.: W. B. Saunders Co., pp. 511–526.

Jacob, Stanley W. and Clarice A. Francone, *Structure and Function in Man*, Philadelphia, Pa.: W. B. Saunders Co., 1965, pp. 463–467.

Kleiner, Israel S. and James M. Orten, *Biochemistry*, 7th ed., St. Louis, Mo.: C. V. Mosby Co., 1966, pp. 712–721.

Metheney, Norma M. and William D. Snively, *Nurses' Handbook of Fluid Balance*, Philadelphia, Pa.: J. B. Lippincott Co., 1967, pp. 28–31.

Nardi, George L. and George D. Zuidema, ed., *Surgery*, 2nd ed., Boston, Mass.: Little, Brown and Co., 1965, pp. 144–145, 157–159.

Pace, Donald, *et al.*, *Physiology and Anatomy*, New York: Thomas Y. Crowell Co., 1965, pp. 24–27.

Ruch, Theodore C. and Harry Patton, ed., *Physiology and Biophysics*, Philadelphia, Pa.: W. B. Saunders, 1965, pp. 899–916.

Shafer, Kathleen N., *et al.*, *Medical-Surgical Nursing*, 3rd ed., St. Louis, Mo.: The C. V. Mosby Co., 1964, pp. 56–57.

Snively, William D., *Sea Within*, Philadelphia, Pa.: J. B. Lippincott Co., 1960, pp. 82–97.

Statland, Harry, *Electrolytes in Practice*, 3rd ed., Philadelphia, Pa.: J. B. Lippincott Co., 1963, pp. 127–154.

Chapter 4

Parenteral Therapy

BEHAVIORAL OBJECTIVES

Upon completion of this chapter, the student will be able to:

Give the three basic objectives for using parenteral therapy.

Write the four main classifications of parenteral solutions and explain the rationale for the use of selected parenteral fluids.

State the methods used in administering parenteral therapy.

State uses of selected parenteral fluids and give examples according to their uses.

Observe clinically patients' physical change when intravenous fluids are being administered.

State the rate of flow of intravenous fluids according to the type of solution, the type of therapy, and the physical condition of the patient.

Calculate the number of drops per minute of intravenous fluid to be absorbed by the body in a specific time.

Describe the physical actions involved in administering parenteral therapy.

Describe the side effects of parenteral therapy and the nursing interventions to meet these effects.

INTRODUCTION

The term parenteral therapy is used to indicate the intake of fluid by a means other than through the alimentary canal. There are various methods of parenteral therapy, but the methods discussed in this chapter are intravenous and subcutaneous (hypodermoclysis) infusions. The basic objectives for using parenteral therapy are to provide maintenance requirements of fluid and electrolytes, to replace previous losses, and to meet concurrent losses. The healthy individual can normally preserve fluid and electrolyte balance, whereas an ill person cannot. When an individual cannot maintain this balance, parenteral therapy is required.

There are various solutions used in preserving fluid and electrolyte balance, one of which is whole blood. About 97% of blood is fluid and, like body water, is comprised of intracellular and extracellular fluids. The plasma portion is typically extracellular fluid and the blood cells contain the intracellular fluid. Whole blood is indicated when there has been severe fluid loss due to hemorrhage, and when the body needs the normal blood constituents, as in anemia.

This chapter discusses the objectives for parenteral therapy, selected solutions commonly used and the rationale for their use, rate of flow for administering solutions, clinical considerations, and nursing intervention for parenteral therapy.

The purpose of this chapter is to help the nurse understand the uses of various parenteral fluids, and the nursing intervention before and during adminsitration. It must be understood that the nurse never administers parenteral fluids without a physician's order. The physician will determine the type and amount of fluid to be given and the time required to administer it. The nurse then computes the hourly requirement in accordance with the physician's order. Her understanding of parenteral therapy and cognizance of physical changes that can occur could be life-saving to the individual. The Nurse Practice Act in many states stipulates that only graduate nurses who have the knowledge, background, and experience can administer intravenous therapy.

1

Refer to the Introduction as needed.

The three basic objectives of administering fluids and electrolytes through parenteral therapy are:

1.
2.
3.

— — — — — — — — — — — — — — —

1. to provide maintenance requirements
2. to replace previous losses
3. to meet concurrent losses

2

Individuals requiring intravenous therapy depend upon this route to meet daily maintenance needs for water, electrolytes, calories, vitamins, and other nutritional substances.

The first objective in administering parenteral therapy is to * _____

_____ .

— — — — — — — — — — — — — —

provide maintenance requirements

3

It is necessary for a patient to have good kidney function while receiving fluids and electrolytes through parenteral therapy. Renal dysfunction could result in electrolyte (retention/excretion).

The electrolyte that can cause cardiac arrest when given in excessive

amounts is _____ .

— — — — — — — — — — — — — —

retention, potassium

4

Maintenance solutions of multiple electrolytes are also helpful in accomplishing the second objective. * _____

_____ of water and electrolytes which may have developed during illness or surgery.

— — — — — — — — — — — — — —

to replace previous (past) losses

5

Fluid and electrolyte losses which can occur due to diarrhea, vomiting, or gastric suction would be an indication of the third objective for administering fluids and electrolytes through parenteral therapy which is

*_____ .

_ _ _ _ _ _ _ _ _ _ _ _ _ _ _

to meet concurrent losses

6

To meet present fluid and electrolyte losses, a replacement solution with an electrolyte content similar to that of the fluid being lost is given in approximately equal amounts to the volume lost.

This type of parenteral therapy would be given to *_____

_____ .

_ _ _ _ _ _ _ _ _ _ _ _ _ _ _

meet concurrent losses

7

Successful fluid and electrolyte therapy often depends on satisfying all three basic objectives which are:

1.
2.
3.

_ _ _ _ _ _ _ _ _ _ _ _ _ _

1. to provide maintenance requirements
2. to replace previous losses
3. to meet concurrent losses

8

Refer to the Introduction as needed. Explain two reasons for administering whole blood.

1.

2.

— — — — — — — — — — — — — —

1. severe fluid loss, e.g., hemorrhage
2. the need for blood constituents, e.g., anemia

9

Blood is composed of _____ % fluid. The plasma contains the (intracellular/extracellular) fluid while the blood cells contain the (intracellular/extracellular) fluid.

— — — — — — — — — — — — — —

97%, extracellular, intracellular

Many of the solutions for parenteral therapy are commercially produced to meet specific types of needs. Table 19 gives a list of commonly used solutions in parenteral therapy and the rationale for their use.

You will be expected to memorize the four major classifications of solutions: protein solutions, plasma expander solutions, hydrating solutions, and replacement solutions. Also you will be expected to memorize the uses for parenteral fluids as stated under rationale and be able to give examples. However, you are not expected to memorize all the solutions and their caloric and electrolyte contents.

Note the tonicity of the solutions as hypotonic, isotonic, or hypertonic. Salt solutions or NaCl are frequently referred to as saline. Normal saline is 0.9% NaCl (sodium chloride).

Study the table carefully because we will be referring to it throughout the chapter. Then go on to the frames that follow, referring back to Table 19 only when necessary.

TABLE 19 Selected Solutions Commonly Used in Parenteral Therapy

Solution	Tonicity	Caloric Value	mEq/L Na^+	mEq/L K^+	Cl^-	Miscellaneous	Rationale for Use of Selected Parenteral Fluids
Protein Solutions							
Aminosol 5%	Iso	175	<10	17	–	Amino acids	Provides protein and fluid for the body. Prevents shock and promotes wound healing.
Aminosol 5% with dextrose 5%	Hyper	345	<10	17	–	Amino acids	Provides protein, calories, and fluid for the body especially helpful for patients who are old and malnourished and for those with hypoproteinemia due to other causes. Not to be used in severe liver damage.
Plasma Expander							
Dextran 6% in saline (dextran 6% and NaCl 0.9%)	Iso	–	154	–	154	–	Colloidal suspensions with physical properties similar to plasma used to increase the plasma volume. Useful in management of burns and shock. They do not replace blood.
Dextran 6% in dextrose 5%	Iso	170	–	–	–	–	
Hydrating Solution							
Sodium chloride 0.45%	Hypo	–	77	–	77	–	Useful for daily maintenance of body fluid but is of less value for replacement of NaCl deficit. Helpful for establishing renal function.
Dextrose 2½% in 0.45% saline	Iso	85	77	–	77	–	Helpful in establishing renal function-urine output.
Dextrose 5% in 0.45% saline	Hyper	170	77	–	77	–	Useful for daily maintenance of body fluids and nutrition, and for rehydration.

Solution	Tonicity	Caloric Value	Na$^+$	K$^+$	Ca^{2+}	Cl$^-$	Lactate	Mg^{2+}	HPO$_4$$^{2-}$	Rationale for Use of Selected Parenteral Fluids
					m mEq/L					
Dextrose 5% in water (50 gram) (dextrose 10% is occasionally used.)	Iso	170	–	–	–	–	–	–	–	Helpful in rehydration and excretory purposes. May cause some sodium loss in the urine. Good vehicle for I.V. potassium.
Replacement Solutions										
Dextrose 5% in saline 0.9%	Hyper	170	154	–	–	154	–	–	–	Replacement of fluid, sodium, chloride, and calories.
Dextrose 10% in saline 0.9%	Hyper	340	154	–	–	154	–	–	–	Replacement of fluid, sodium, chloride, and calories.
Lactated Ringer's	Iso	9	130	4	3	109	28	–	–	This solution resembles the electrolyte structure of normal blood serum-plasma. Potassium present is not sufficient for daily potassium requirement.
Dextrose 5% in lactated Ringer's	Hyper	179	130	4	3	109	28	–	–	Same contents as lactated Ringer's plus calories.
Ringer's solution	Iso	–	147	4	5	156	–	–	–	Does not contain the lactate which can be harmful to individuals who cannot metabolize lactic acid.
Dextrose 5% in Ringer's solution	Hyper	170	147	4	4	155	–	–	–	Same contents as Ringer's solution plus calories.
M/6 sodium lactate	Iso	55	167	–	–	–	167	–	–	Supplies sodium without the chloride. The lactate has some caloric value and is metabolized to CO$_2$ for excretion or increases bicarbonate in alkalosis.

TABLE 19 (continued)

Solution	Tonicity	Caloric Value	Na⁺	K⁺	Ca²⁺	Cl⁻	Lactate	Mg²⁺	HPO₄²⁻	Rationale for Use of Selected Parenteral Fluids
			— mEq/L							
Normal saline (0.9%)	Iso.	–	154	–	–	154	–	–	–	Restores extracellular fluid volume and replaces sodium chloride deficit.
Hypertonic saline 5% NaCl	Hyper	–	855	–	–	855	–	–	–	Helpful in salt-depleted hyponatremia. It raises the osmolality of the blood. Helpful in eliminating water excess from cells.
Ionosol B with dextrose 5%	Hyper	178	57	25	–	49	25	5	13	Useful in treating patients requiring a polyionic parenteral replacement, e.g., with alkalosis due to vomiting, diabetic acidosis, and fluid losses due to burns or stress, and post-operative dehydration.
Ionosol D–CM with dextrose 5%	Hyper	186	138	12	5	108	50	3	–	For replacement of electrolyte losses of duodenal fluid through intestinal suction, biliary or pancreatic drainage, and to correct mild acidosis.

Source: Selected portions from Abbott Laboratories, *Wall Chart, Intravenous and Other Solutions*, October, 1968; Harry Statland, *Fluid and Electrolytes in Practice*, Philadelphia, Pa.: J. B. Lippincott Co., 1963.

10

The four main classifications of parenteral solutions are:
1.
2.
3.
4.

– – – – – – – – – – – – – –

1. protein solution
2. plasma expander
3. hydrating solutions
4. replacement solutions

11

The solution with physical properties similar to plasma is the plasma ex-

pander _____ .

The plasma expander solution is useful in the management of _____

and _____ . It (does/does not) replace blood.

– – – – – – – – – – – – – –

dextran 6% or dextran 6% in saline, burns and shock, does not

12

The solution resembling the electrolyte structure of plasma is * _____

_____ .

– – – – – – – – – – – – – –

lactated Ringer's or Ringer's solution

13

In what way does lactated Ringer's solution and dextran solution resemble plasma?

Lactated Ringer's. _____ .

Dextran. _____ .

_ _ _ _ _ _ _ _ _ _ _ _ _ _ _

lactated Ringer's—electrolyte structure
dextran—colloidal solution

14

Hydrating solutions are helpful in daily maintenance of body fluid, rehydration, and in establishing effective renal output.
    Indicate which of the following are hydrating solutions:
( )   a. Dextrose 2 1/2% in 0.45% saline
( )   b. Ringer's solution
( )   c. Dextrose 5% in water
( )   d. Sodium chloride 5%
( )   e. Dextrose 5% in 0.45% saline
( )   f. Sodium chloride 0.45%

a. X   d. –
b. –   e. X
c. X   f. X

15

Potassium is often administered by diluting an ampul or two of potassium chloride in solution. A good vehicle for this I.V. potassium is the

hydrating solution, *_____ .

_ _ _ _ _ _ _ _ _ _ _ _ _ _ _

1000 cc or ml of dextrose 5% in water

16

Replacement solutions are useful for replacing fluid, calories, and electrolyte deficits due to injury or illness.

Indicate which of the following solutions are considered replacement solutions:

( ) a. Sodium chloride 0.45%
( ) b. Dextrose 5% in normal saline
( ) c. Lactated Ringer's
( ) d. Ringer's
( ) e. Dextrose 5% in water
( ) f. M/6 sodium lactate
( ) g. Hypertonic (3% or 5%) saline
( ) h. Dextrose 5% in 0.45% saline
( ) i. Multiple electrolyte or polyionic solutions (Ionosol B or D-CM solutions)

----

a. −   f. X
b. X   g. X
c. X   h. −
d. X   i. X
e. −

17

The solution that contains sodium and not chloride is *_____

_____.

M/6 sodium lactate

18

Hypertonic saline (3% or 5%) is used in treatment of (hypernatremia/hyponatremia).

This solution will raise the _____ of the blood.

----

hyponatremia, osmolality

19

The multiple electrolyte solution most helpful in treating severe vomiting, diabetic acidosis, post-operative dehydration, and fluid loss due to

burns is *_____ .

– – – – – – – – – – – – – – – –

Ionosol B with dextrose 5%

20

The multiple electrolyte solution most helpful in replacement of electro-

lyte losses from the gastrointestinal tract is _____ .

Can you explain why? _____

_____ .

– – – – – – – – – – – – – – – –

Ionosol D-CM
Electrolyte content of the solution is similar to the gastrointestinal
juice.

21

Do you think lactated Ringer's solution would be administered in a case

of potassium deficit? Why or why not? _____

_____ .

– – – – – – – – – – – – – – –

No! The amount of potassium is not sufficient in replacing potassium
deficits.

22

Dextran 6% in saline is a *_____ , because its physi-

cal properties are similar to those of _____ .

– – – – – – – – – – – – – –

plasma expander, plasma

23

Name three purposes for the use of hydrating solutions:
1.
2.
3.

— — — — — — — — — — — — — — —

1. daily maintenance
2. rehydration
3. effective renal output

24

Replacement solutions are useful for replacing _____

_____ .

— — — — — — — — — — — — — — —

fluid, calories, and electrolyte deficits

METHODS FOR ADMINISTRATION OF PARENTERAL THERAPY

25

The four routes available for parenteral therapy are: the intravenous, the intramuscular, the subcutaneous, and the intramedullary. Individuals with severe disturbance of fluid and electrolyte balance are most likely to require intravenous therapy. Occasionally, subcutaneous therapy is desired in place of intravenous therapy.

The preferred route for administering parenteral therapy for fluid and

electrolyte disturbances is _____ .

— — — — — — — — — — — — — — —

intravenously

Diagram 10 shows the method for administering subcutaneous parenteral therapy. Subcutaneous refers to the fatty tissue beneath the skin surface. The needle is inserted at a 45° angle through the skin. The thighs are choice areas for insertion. The posterior aspect of the chest is a choice area for infants.

The subcutaneous parenteral therapy is known as clysis or hypodermoclysis.

Refer to Diagram 10 as needed.

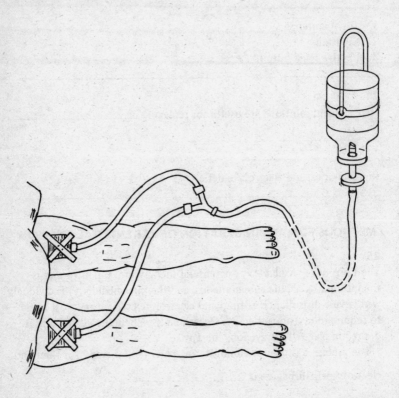

Diagram 10. Hypodermoclysis

26

The types of solution used for clysis are <u>hypotonic</u> and isotonic. Hypertonic solutions are warned against for they can draw fluid into the area of clysis, diminishing the plasma volume, which would lead to circulatory collapse.

Two names for subcutaneous therapy are _____ and

_____ .

– – – – – – – – – – – – – – –

clysis, hypodermoclysis

27

Types of solutions desired for subcutaneous therapy are _____

and _____ .

– – – – – – – – – – – – – –

hypotonic and isotonic

28

What type of solution should <u>not</u> be used in subcutaneous therapy?

_____ . Why? _____

_____ .

The types of isotonic solutions frequently used in clysis are normal saline (0.9%) and dextrose 2½% in half-strength saline (0.45%).

When giving dextrose 5% in water (isotonic solution) by hypodermoclysis, NaCl will diffuse to the clysis area from the surrounding tissues and blood. This can lead to (hyponatremia/hypernatremia).

The desired isotonic solutions for clysis (hypodermoclysis) are

*_____ and _____

_____ .

– – – – – – – – – – – – – – –

hypertonic solution, It will draw fluid into the area of clysis and diminish the intravascular fluid volume., hyponatremia, dextrose 2½% in 0.45% saline and normal saline (0.9%)

Review

Q. 1
The three basic objectives of parenteral fluid and electrolyte therapy are:
1.
2.
3.

_ _ _ _ _ _ _ _ _ _ _ _ _

1. to provide maintenance requirements
2. to replace previous losses
3. to meet concurrent losses

Q. 2
To meet present fluid and electrolyte losses, a _____ solution would be recommended.

_ _ _ _ _ _ _ _ _ _ _ _ _

replacement

Q. 3
In giving parenteral fluid with electrolytes, the kidney function is extremely important. Why? _____

_____ .

_ _ _ _ _ _ _ _ _ _ _ _ _

Renal dysfunction could result in electrolyte retention.

Q. 4
The two common conditions in which blood is given are for_____

_____ and _____ .

_ _ _ _ _ _ _ _ _ _ _ _ _

hemorrhage and anemia

Q. 5

Blood is frequently given intravenously to:

1.
2.

– – – – – – – – – – – – – – –

1. replace severe fluid loss
2. provide the body with needed blood constituents

Q. 6

The plasma expander solution is useful in the management of _____

and _____ .

– – – – – – – – – – – – – – –

burns and shock

Q. 7

The solution resembling the electrolyte structure of plasma is *_____

_____ .

– – – – – – – – – – – – – – –

lactated Ringer's solution

Q. 8

Hypertonic saline (3% or 5%) will raise the _____ of the
blood.

– – – – – – – – – – – – – – –

osmolality

Q. 9

The polyionic parenteral solutions are helpful in replacement of electro-

lyte losses from the *_____ and in treating dia-

betic acidosis, postoperative dehydration, and *_____ .

– – – – – – – – – – – – – – –

gastrointestinal tract, fluid loss due to burns

Q. 10

For treatment of fluid and electrolyte imbalance, how would parenteral therapy usually be administered? _____.

_ _ _ _ _ _ _ _ _ _ _ _ _ _ _

intravenously

Q. 11

Name two types of solution used for subcutaneous therapy. _____

_____ and _____.

_ _ _ _ _ _ _ _ _ _ _ _ _ _ _

dextrose 2½% in 0.45% NaCl and normal saline

Q. 12

Match the following solutions according to their classification.

_____ 1. Dextrose 5% in water

_____ 2. Lactated Ringer's solution

_____ 3. Aminosol

_____ 4. Sodium chloride 0.45%

_____ 5. Sodium chloride 0.9%

_____ 6. Dextran 6% in saline

_____ 7. Dextrose 5% in normal saline (0.9%)

a. Protein solution
b. Plasma expander
c. Hydrating solution
d. Replacement solution

_ _ _ _ _ _ _ _ _ _ _ _ _ _ _

1. c   5. d
2. d   6. b
3. a   7. d
4. c

Clinical Consideration

29

The first step in treatment of fluid and electrolyte disturbances should be the reconstitution of the extracellular fluid and blood volume. This can best be accomplished by isotonic saline (normal saline), lactated Ringer's, or M/6 sodium lactate.

The isotonic solutions that can be used in the reconstitution of the extracellular fluid and blood volume are:

1.
2.
3.

– – – – – – – – – – – – – – – –

1. normal saline
2. lactated Ringer's
3. M/6 lactated sodium

30

Do you think the volume of urine would increase or decrease following

the administration of these isotonic solutions? _____ Why?

_____ .

– – – – – – – – – – – – – – –

Increase., There would be more fluid in the body to be excreted.

31

Hematocrit (volume of red blood cells in proportion to the intravascular fluid) reading would be one indication of gain or loss of the blood volume. An increased hematocrit reading could indicate an intravascular

fluid loss. Do you know the reason? _____ .

Explain. _____

_____ .

– – – – – – – – – – – – – – –

Yes? Good; No? Intravascular fluid loss would increase the number of red blood cells in proportion to the fluid.

32
Concentration of red blood cells is known as hemoconcentration.
 A high hematocrit reading could be an indication of:
( ) a. dehydration
( ) b. overhydration

_ _ _ _ _ _ _ _ _ _ _ _ _ _ _

a. X
b. —

33
Giving whole blood or plasma will dilute the hemoconcentration, lower
the hematocrit, raise the blood pressure, and establish renal flow.
 How do you think whole blood would dilute the hemoconcentration?

_____

_____.

_ _ _ _ _ _ _ _ _ _ _ _ _ _ _

There would be a slight increase in the intravascular fluid since 55%
whole blood is plasma; however, plasma would be a better choice in
dilution.

34
For best results, whole blood should be given following initial dilution
by an isotonic solution.
 Explain why you think an isotonic solution should be given first for

hemoconcentration before blood. _____

_____.

_ _ _ _ _ _ _ _ _ _ _ _ _ _ _

It will aid in immediate dilution and will also prevent clotting of blood.

35

The isotonic solutions helpful in reconstitution of the extracellular

fluid and blood volume are whole blood, plasma, _____,

_____, and _____.

_ _ _ _ _ _ _ _ _ _ _ _ _ _ _

normal saline, lactated Ringer's, and M/6 sodium lactate

36

Whole blood given will:

( ) a. raise blood pressure
( ) b. lower blood pressure
( ) c. increase renal flow
( ) d. decrease renal flow

_ _ _ _ _ _ _ _ _ _ _ _ _ _ _

a. X   c. X
b. –   d. –

37

Dextrose 5% in water, although isotonic when administered, has little
effect in increasing blood volume since the sugar is utilized rapidly.

Dextrose in water (will increase/will not increase) the blood volume.

_ _ _ _ _ _ _ _ _ _ _ _ _ _ _

will not increase

38

As blood volume is being restored, attention is directed to the osmolal
changes, correcting the hypertonicity or the hypotonicity. Dextrose 5%
in water is effective in correcting water deficit.

Do you think dextrose 5% would be administered to a patient with

hyponatremia? _____. Why or why not? _____

_____.

_ _ _ _ _ _ _ _ _ _ _ _ _ _ _

NO. Body salt would be superdiluted.

39

A gross water deficit can cause kidney function impairment. With a decreased urine output resulting from renal impairment, metabolic acidosis can occur since most of the hydrogen ions cannot be excreted. There would be a retention of hydrogen ions.

A gross water deficit can also result in cell damage, causing hypokalemic alkalosis. The plasma CO_2 content becomes markedly elevated, resulting in alkalosis, and the potassium leaves the cell due to cell damage.

With a decrease in cellular potassium, there would be a (decrease/increase) in plasma bicarbonate. Loss of potassium is accompanied by

a loss of _____.

– – – – – – – – – – – – – – – –

increase, chloride

40

A gross water deficit can cause * _____ and

_____.

– – – – – – – – – – – – – – –

metabolic acidosis, hypokalemic alkalosis

41

In severe cases of vomiting, metabolic alkalosis results from loss of chloride and potassium. Metabolic acidosis results in loss of bicarbonate and other electrolytes.

The solution effective in correcting metabolic alkalosis with hypokalemia would be sodium (chloride/bicarbonate) with potassium added.

The solution effective in correcting metabolic acidosis would be

* _____.

– – – – – – – – – – – – – – –

chloride, sodium bicarbonate

42

In treatment of respiratory alkalosis, replacement of bicarbonate is not generally necessary.

In treatment of respiratory acidosis, the chloride deficit would be present; therefore, solutions of sodium chloride would be given.

When necessary, the treatment of respiratory alkalosis (blowing off of CO_2) would be sodium _____ .

The solution effective in correcting respiratory acidosis (retention of CO_2) would be * _____ .

— — — — — — — — — — — — — — — —

bicarbonate, sodium chloride

43

The physician will determine the need for parenteral therapy by checking on the hematocrit, the B.U.N. (blood urea nitrogen), and the serum electrolytes.

Hematocrit is the volume of red blood cells in proportion to the

* _____ .

What might a high hematocrit indicate? _____ .

Why? _____ .

— — — — — — — — — — — — — —

intravascular fluid, hemoconcentration or dehydration
The proportion of red blood cells is greater than the proportion of fluid.

44

Blood urea nitrogen (B.U.N.) is a serum test to determine the amount of urea, a by-product of protein metabolism, remaining in the blood which is normally excreted by the kidneys.

An increased B.U.N. frequently means poor renal function.

The kidneys will normally (retain/excrete) urea. An increased B.U.N.

could mean _____.

_ _ _ _ _ _ _ _ _ _ _ _ _ _ _

excrete, poor renal function

45

The third method for determining the need for parenteral therapy is

through checking *_____.

To determine whether or not parenteral therapy is necessary, a physician will check:

1.
2.
3.

_ _ _ _ _ _ _ _ _ _ _ _ _ _ _

serum electrolytes
1. hematocrit
2. B.U.N.
3. electrolytes

46

Fluids and electrolytes for maintenance therapy should be continuous for 24 hours and rescheduled on a day-to-day basis.

Normally, infusion of fluids results in prompt excretion of any excess water and electrolyte. This response defends the body against

*_____ in water and electrolyte balance.

How should fluids and electrolytes for maintenance therapy be scheduled? _____.

Why? _____.

_ _ _ _ _ _ _ _ _ _ _ _ _ _ _ _

excess or significant alterations
CONTINUOUS on a day-by-day basis. Why? This defends the body against significant alterations.

47

If a patient receives his full 24-hour maintenance parenteral therapy in 8 hours, two-thirds of the water and electrolyte will be in (excess/deficit) of current needs and, normally, a large quantity of the excess maintenance fluid will be (excreted/retained).

_ _ _ _ _ _ _ _ _ _ _ _ _ _ _

excess, excreted

48

Tolerance for water and electrolytes is limited for the very ill patients, patients following surgery, aged patients, small children, and infants.

Rapid administration of maintenance fluids which exceed physiologic tolerance can cause hyponatremia, pulmonary edema, which is accumulation of fluid in the chest, and other complications.

Maintenance parenteral therapy should be administered over a period

of _____.

What are the two possible results from rapid administration of maintenance fluids? _____ and _____.

– – – – – – – – – – – – – – – – –

24 hours, hyponatremia and pulmonary edema

49

Hypertonic solutions, those having greater osmotic activity than the body fluids, should be injected not faster than 3 to 4 ml per minute or as ordered by the physician. Three to 4 ml is equivalent to 45-60 drops (gtts), Abott fluid set, or 30–40 gtts, Baxter fluid set.

List at least five hypertonic solutions to which this might apply. Refer back to Table 19 if needed.
1.
2.
3.
4.
5.

– – – – – – – – – – – – – – – –

1. Dextrose 5% in normal saline
2. Dextrose 10%
3. Dextrose 10% in normal saline
4. Dextrose 5% in lactated Ringer's solution
5. Hypertonic saline
6. Multiple electrolyte solutions

Table 20 outlines the three types of therapy frequently employed with intravenous administration. The table includes the desired amount of solution and the rate of flow for each type of therapy. The symbol ml or milliliter has the same equivalence as cc or cubic centimeter. The symbol for drops is gtts.

The asterisks refer to the number of drops equivalent to one milliliter, according to the type of intravenous sets used. The single asterisk is the Abbott set: 15 gtts = 1 ml and the double asterisks refer to the Baxter set: 10 gtts = 1 ml.

Take several minutes to study this table carefully. Memorize the equivalent drops per milliliter for both intravenous sets. Know the three types of therapy, the recommended dosage, and the rate in which the fluid should be administered. Refer to this table as needed.

TABLE 20   Rates of Intravenous Administration

Type of Therapy	Amount of Solution Desired	Rate of Flow
Maintenance therapy	1500–2000 ml	2 ml/minute (30 gtts,* 20 gtts**) if given over 24 hours. 4 ml/minute (60 gtts) in two infusions.
Replacement with maintenance therapy	2000–3000 ml	3 or 4 ml/minute (45–60 gtts,* 30–40 gtts**). (Depends upon the individual and his physician.)
Hydration therapy	1000–3000 ml (Isotonic)	8 ml/minute (120 gtts,* 80 gtts**) for the first 45 minutes. If urinary output is not established, then 2 ml/minute the next hour. After urinary output is established, continue as in maintenance therapy.

These guidelines may be adapted to individual circumstances. The physician sets the 24-hour requirements and the registered nurse computes 1-hour requirements from this. The amount of solution to be administered and the rate of flow can vary greatly with the very sick, the aged individual, the small child, and the infant.
*Abbott set = 15 gtts per 1 ml.
**Baxter set = 10 gtts per 1 ml.

50

The names of the three types of parenteral therapy are:
1.
2.
3.

– – – – – – – – – – – – – –

1. maintenance therapy
2. replacement therapy
3. hydration therapy

51

The type of solution most frequently used in hydration therapy is:
( ) a. hypertonic
( ) b. isotonic
Once urinary output is reestablished, then (maintenance/replacement) therapy is started.

– – – – – – – – – – – – –

a. –
b. X
maintenance

52

The amount of solution to be administered and the rate of flow can vary greatly for what individuals? _____ , the _____

_____ , the _____ , and the _____ .

– – – – – – – – – – – – – –

a very sick person, the aged individual, small child, infant

53

The physician sets the 24-hour requirements and the registered nurse computes *_____ .

– – – – – – – – – – – – – –

1-hour requirements from this

54

The number of drops per ml of the commercial administration sets for intravenous fluids will vary according to the manufacture of the set. Some sets as manufactured by a given company are calculated 10 gtts (drops) per 1 ml (cc) while sets of other manufacturers are calculated at 15 gtts per 1 ml.

The drops per ml can vary according to *_____

_____ .

– – – – – – – – – – – – – – – –

the manufacturer of the intravenous sets

55

Your patient is to receive 1000 ml of 5% dextrose in H_2O in 8 hours. Calculate the number of drops per minute according to the Abbott set.

Example A: $1000 \div 8 = 125$ ml (cc) in 1 hour

125 ml X 15 drops per ml = _____ drops in 1 hour

1875 drops ÷ 60 minutes = _____ drops per minute
(1 hour)

– – – – – – – – – – – – – – – –

1875 drops (gtts), 30–32 drops (gtts)

56

Your patient is to receive 1000 ml of 5% dextrose in 0.45% saline in 12 hours. Calculate the number of drops per minute according to the Baxter set.

Example B: $1000 \div 12 = 83$ ml (cc) in 1 hour

83 ml X 10 drops per ml = _____ drops in 1 hour

830 drops ÷ 60 minutes = _____ drops per minute

– – – – – – – – – – – – – – – –

830 drops (gtts), 13–14 drops (gtts)

57

Here is another problem in intravenous therapy. Your patient's intravenous order is 500 ml of 5% dextrose in normal saline to be administered in 4 hours using the Baxter set. Calculate the number of drops per hour and per minute. Work space is provided.

   Example C:

_ _ _ _ _ _ _ _ _ _ _ _ _ _

125 ml per hour, 1250 drops (gtts) per hour, 20–21 drops (gtts) per minute

Review

Q. 1

The first two steps in treatment of fluid and electrolyte disturbances

should be the reconstitution of the *_____ and

_____.

_ _ _ _ _ _ _ _ _ _ _ _ _ _

extracellular fluid and blood volume

Q. 2

A high hematocrit reading would be indicative of a _____
blood volume.

_ _ _ _ _ _ _ _ _ _ _ _ _

low, or decreased

Q. 3

How would the administration of whole blood or plasma affect the

hematocrit? _____

_____ .

— — — — — — — — — — — — — —

The hematocrit would be lowered, the blood pressure would increase, and the renal flow would be reestablished or increased.

Q. 4

A gross body water deficit can result in cell _____

causing a hypokalemic alkalosis. Why? _____

_____ .

— — — — — — — — — — — — — —

damage, or destruction, CO_2 content becomes elevated and the potassium leaves the damaged cells

Q. 5

A gross body water deficit resulting in kidney function impairment can

cause metabolic acidosis. Why? _____ .

— — — — — — — — — — — — — —

Retention of the hydrogen ions

Q. 6

To correct metabolic alkalosis, you might administer *_____

_____ , while correction of metabolic acid-

osis would be to administer *_____ .

— — — — — — — — — — — — — —

sodium chloride with potassium added, sodium bicarbonate

Q. 7

If an individual receives large quantities of fluid in a short time, the urinary output of fluid and electrolyte will be _____ .

– – – – – – – – – – – – – – –

increased

Q. 8

What are the two possible results from rapid administration of parenteral fluids which exceeds physiological tolerance? _____ and _____ .

– – – – – – – – – – – – – –

hyponatremia and pulmonary edema

Q. 9

Rapid administration of maintenance fluids is not indicated for what individuals? _____ , _____ , _____ , and _____ .

– – – – – – – – – – – – – – –

very sick patient, aged individual, small child, infant

Q. 10

The physician will determine the need for parenteral therapy by checking on the:
1.
2.
3.

– – – – – – – – – – – – – – –

1. hematocrit
2. B.U.N.
3. serum electrolytes

NURSING INTERVENTION

58

The nurse has many responsibilities in parenteral therapy, especially when there are serious fluid and electrolyte imbalances.

The nurse must keep careful records of what the patient is receiving and assume responsibility for accurate record of intake and output of the individual.

The nurse must know what parenteral fluids her patient is receiving

and keep an accurate record of *_____.

_ _ _ _ _ _ _ _ _ _ _ _ _ _ _ _

intake and output

Table 21 discusses the nurse's responsibility in the management of parenteral therapy. The physician orders the parenteral fluids, and the nurse must have the knowledge and understanding of parenteral therapy and the various solutions used in order to manage parenteral therapy.

This table is divided into two parts. Part I describes the physical actions involved in administering parenteral therapy. It includes the clinical considerations, nursing interventions, and rationale. Study Part I carefully and go immediately to the frames that follow.

Part II describes the side effects of parenteral therapy. It includes the clinical considerations, adverse reactions, and nursing interventions. Study Part II carefully and go immediately to the frames that follow.

Refer to this table whenever it is necessary.

TABLE 21 The Nurse's Responsibility in the Management of Parenteral Therapy
Part I. Physical Actions Involved in Administering Parenteral Therapy

Clinical Considerations	Nursing Interventions	Rationale
Solution for intravenous fluid	Use solutions that are at room temperature and not cold from the refrigerator.	Solutions given at room temperature are rapidly absorbed by the patient.
		Very cold solution can result in vasoconstriction.
		Fluids given slowly will mix with the blood or fluid in the tissue spaces and will be quickly brought to body temperature.
Flow rate	Check the types of solutions patients are receiving.	Knowledge of the tonicity of fluids will aid in determining the rate of flow. The rate for hypertonic solutions should be slower than that of isotonic solutions.
	Observe the drip counter and regulate accordingly.	
		Regulation of intravenous fluids is important in the prevention of overhydration.
Injection site	Choose the distal portion of the arm for intravenous therapy.	The distal portion of the arm is preferred rather than the elbow in order to avoid the necessity for immobilizing the elbow and to avoid the possibility of infiltration into the interstitial spaces.
	Observe the injection site for infiltration.	By checking the appearance of the injection site, infiltration into the interstitial spaces could be observed and damage to the surrounding tissues could be prevented.
Intake and output	Keep an accurate intake and output record.	Accurate intake and output records will indicate fluid and electrolyte balance or imbalance. It will determine the degree of fluid retention and ensure that losses do not exceed replacement.
	Check hourly output for those with known fluid and electrolyte imbalance and for those receiving potassium.	

TABLE 21 Part I (continued)

Clinical Considerations	Nursing Interventions	Rationale
	Check that the patient is not becoming overloaded with liquids—orally and parenterally.	An hourly check on the output can help in correction of fluid and electrolyte imbalance before it becomes irreversible.
Intravenous needles	Note when the parenteral therapy began and the type of needle used.	Intravenous needles should not be left in site for periods longer than 24–48 hours without producing a local inflammatory reaction and/ or painful thrombophlebitis.
	Change the injection site every 24–48 hours when using a 20–21 gauge needle or every 72–96 hours when using polyethylene tubing.	If a polyethylene tubing (intracath) is used in the vein, this should not remain longer than 4 days without producing a phlebitis or a thrombophlebitis.
Vital signs	Check the patient's blood pressure, temperature, pulse, and respiration for any adverse reaction while the patient is receiving parenteral therapy.	Identification of abnormal vital signs could eliminate complications.
	Identify adverse reactions according to the kind of parenteral therapy used.	Any adverse reaction during intravenous therapy could change the vital signs, e.g., in blood transfusion, a rise in temperature and pulse rate could indicate blood incompatibility.

59
The rationale for giving intravenous solutions at room temperature and not cold solutions from the refrigerator would be:
1.
2.
3.

– – – – – – – – – – – – – –

1. Room temperature solutions would be absorbed more rapidly.
2. Cold solutions can result in vasoconstriction.
3. Warm fluids given slowly will mix quicker with the blood and interstitial tissues.

60
Explain why the nurse needs to know the tonicity of the intravenous

fluid. _____ .

– – – – – – – – – – – – – –

It will aid in determining the rate of flow.

61
The flow rate for hypertonic solutions should not exceed _____
ml per minute.

– – – – – – – – – – – – – –

3 to 4 ml

62
The recommended area for intravenous therapy is * _____

_____ .

  The rationale for using the distal portion of the arm instead of the

elbow is to avoid * _____

_____ .

– – – – – – – – – – – – – –

the distal portion of the arm, elbow immobilization and the possibility of infiltration.

63

The rationale for observing for subcutaneous infiltration is to prevent

*_____ .

— — — — — — — — — — — — — — —

damage to the surrounding tissue

64

How often should a nurse check the urinary output for individuals with

renal dysfunction who are receiving electrolyte supplements? _____ .
  Individual having poor urinary output and receiving potassium should

be watched for potassium intoxication which could lead to *_____

_____ .

— — — — — — — — — — — — — — —

hourly, cardiac arrest

65

The nursing intervention in relation to intravenous needles is to know

when the *_____ began and the type of _____

used.

— — — — — — — — — — — — — — —

parenteral therapy, needle

66

Intravenous needles left in the same site for periods longer than 24–48

hours could cause a *_____ .

— — — — — — — — — — — — — — —

local inflammatory reaction or thrombophlebitis

67
Why should the nurse check vital signs during parenteral therapy? ____

_____.

Identification of abnormal vital signs while the patient is receiving

parenteral fluids could eliminate many *_____.

_ _ _ _ _ _ _ _ _ _ _ _ _ _ _

To determine any adverse reactions, serious complications

68
Explain what changes in the vital signs could indicate blood incompati-

bility. _____.

_ _ _ _ _ _ _ _ _ _ _ _ _ _ _

Rise in temperature and pulse rate

TABLE 21 Part II. Side Effects of Parenteral Therapy

Clinical Considerations	Adverse Reactions	Nursing Interventions
Varying Flow Rate According to Specific Solutions. Hydrating solutions	Large quantities of hydrating solutions could result in over-hydration.	Note that hydrating solutions are mainly isotonic and aid in restoring urinary output.
		Note that isotonic solutions can replace fluid loss quickly since they do not increase osmotic pressure.
		Regulate the rate of hydrating solutions to run fast at first and then at a slower rate thereafter unless over-hydration is present.
		Identify symptoms of over-hydration, e.g., dyspnea, coughing, and cyanosis.

TABLE 21   Part II (continued)

Clinical Considerations	Adverse Reactions	Nursing Interventions
Protein solutions, also known as amino acid solutions	Reactions to amino acid solutions frequently occur from administering the solution too fast. These reactions include nausea, feeling of warmth, and flushing.	Note these solutions are helpful in combating hypoproteinemia and conditions due to malnutrition. Reguate the rate of amino acid solutions to run slowly at first and later to run faster if patient can tolerate the solution. Observe and identify adverse reactions to amino acid solutions.
Dextrose solutions	Isotonic dextrose can cause a rise in CNS fluid pressure. Symptoms of increased pressure, also referred to as cerebral edema, include: slow pulse rate, high blood pressure, and headaches.	Note that dextrose cannot be metabolized more rapidly than 1 gram/kg (kilogram) or 4.5 gram per 10 pounds of body weight per hour. Regulate the rate of flow not to exceed the metabolizing rate, especially for patients who are very sick or with head injuries. Observe to see if patients exhibit symptoms of increased CNS fluid pressure or cerebral edema, especially patients with head injuries.
Hypertonic solutions	Hypertonic solutions when administered too fast, can cause sclerosis of a recipient's vein or can cause thrombophlebitis. Also hypertonic solutions can pull fluid from the brain lowering the CNS fluid pressure. Symptoms are those similar to shock: fast pulse rate, low blood pressure, and restlessness.	Note which intravenous fluids are hypertonic. Regulate the rate of hypertonic solutions to run no faster than 3–4 ml/minute. Observe the patient for symptoms of shock. Frequently, hypertonic solutions are used in treating cerebral edema. If too much is administered, dehydration or loss of water occurs and later shock.

TABLE 21 Part II (continued)

Clinical Considerations	Adverse Reactions	Nursing Interventions
Overhydration by parenteral therapy	An increase in the respiratory rate, the onset of coughing and/or the development of cyanosis would suggest an overhydration of the patient's circulation with transudation of fluid from the blood into the aveoli of the lungs, thus resulting in pulmonary edema.	Note the clinical diagnosis and physical changes of patients receiving parenteral therapy.
		Observe changes in rate and depth of respiration. Dyspnea, difficulty in breathing, accompanies pulmonary edema or fluid.
		Observe for a constant and an irritated cough, which can result from pulmonary fluid.
		Observe for signs of cyanosis, resulting from a lack of pulmonary ventilation, depriving tissues of O_2.
		Observe individuals with cardiac, renal, and/or liver diseases receiving parenteral therapy, for signs of overhydration, including edema, which is increased body water.
		Regulate the flow rate according to the patient's clinical condition and the solution used.
		Discontinue injection when discovering adverse signs.
Increased fluid intake	Patients receiving large quantities of oral fluid along with large quantities of intravenous fluid can overload the intravascular system.	Monitor the fluid intake, oral and parenteral, on postoperative patients of one day, the very sick, the cardiac, and patients with chronic renal and liver diseases.
Inadequate urinary output	Patients having poor urinary output and receiving potassium could experience cardiac arrest.	Check the patient's urinary output before and during the administration of parenteral fluids having potassium.

TABLE 21   Part II (continued)

Clinical Considerations	Adverse Reactions	Nursing Interventions
Electrolyte Imbalance		
Sodium retention	Sodium solutions given in excess quantities can result in edema due to the retention of sodium and fluid. Tachycardia with a fall in blood pressure could indicate hypernatremia.	Observe signs and symptoms of developing edema.
Potassium deficit	Potassium deficit could result from surgery, trauma, injury, stress, dehydration, diarrhea, vomiting, anorexia, and such, along with dilution due to excessive parenteral fluids. Dizziness, muscular weakness, arrhythmia could indicate hypokalemia.	Observe signs and symptoms of potassium deficit or hypokalemia.
Potassium excess	Potassium excess can be most hazardous, leading to cardiac arrest. Concentration of potassium in 1000 ml of fluid exceeding 40 mEq/L or two 20% ampules administered in less than 4 hours could result in potassium intoxication and later death. Nausea, abdominal cramps, tachycardia, followed by bradycardia could indicate hyperkalemia.	Note the "safe" range for using concentrated potassium salts. Nurses preparing concentrated potassium chloride (KCl) ampules to be diluted in intravenous solution should be cognizant of this responsibility and to realize that an error in dilution could be fatal. Check to determine the concentration of potassium does not exceed 40 mEq/l per 1000 ml. 30 mEq/L is preferred to avoid potassium intoxication. Administer potassium via intravenous solutions over a period of no less than 4 hours and 3 ml per minute unless otherwise ordered. Observe for signs and symptoms of potassium excess or hyperkalemia.

69

Hydrating solutions help in restoring * _____ .

_ _ _ _ _ _ _ _ _ _ _ _ _ _ _

urinary output

70

Isotonic solutions can replace fluid less quickly because * _____

_____ .

Large quantities of hydrating solutions could result in _____ .

_ _ _ _ _ _ _ _ _ _ _ _ _ _ _

they do not create an increase in osmotic pressure, overhydration

71

Amino acid solutions are helpful in combating hypoproteinemia and

* _____ .

_ _ _ _ _ _ _ _ _ _ _ _ _ _ _

conditions caused by malnutrition

72

Amino acid solutions should run slowly at first. Why? _____

_____ .

Some adverse reactions to amino acid solutions are _____ ,

_____ , or _____ .

_ _ _ _ _ _ _ _ _ _ _ _ _ _ _

To see if the patient can tolerate the solution.
nausea, feeling of warmth, or flushing

73

Dextrose is metabolized at a rate of _____ per 10 pounds
per hour.

_ _ _ _ _ _ _ _ _ _ _ _ _ _ _

4.5 gram

74

Dextrose solutions for intravenous therapy are prepared in two forms—5% and 10%. The 5% dextrose means there are 5 grams in 100 ml. If the bottle contains 1000 ml of 5% dextrose, then how many grams would be in this bottle? _____ .

– – – – – – – – – – – – – – – –

50 grams: 5 grams in 100 ml

$5 \times 10 = 50$ grams

or by ratio: $5 : 100 :: x : 1000$

or by fraction: $\dfrac{5}{100} = \dfrac{x}{1000}$

$100x = 5000$

$x = 50$ grams

75

A patient weighs 130 pounds and is to receive 1000 ml of intravenous 5% dextrose in water. How many grams of dextrose should your patient receive in 1 hour in order for the dextrose to be properly metabolized?

$130 \div 10 =$ _____

4.5 grams $\times$ _____ = _____

If your patient is to receive 58.5 grams, then how much 5% dextrose solution should your patient receive? _____ .

By fraction:

$\dfrac{5}{100} = \dfrac{58.5}{x}$

$5x = 5850.0$

$x =$ _____

– – – – – – – – – – – – – – – –

13, 13 = 58.5 grams
1170 ml, 1170 ml

76

Another problem:

Your patient weighs 150 pounds and is to receive 1000 ml of 10% dextrose in water. How much solution should your patient receive?

_____.

> By fraction:
>
> $$\frac{10}{100} = \frac{67.5}{x}$$
>
> $10x = 6750.0$
>
> $x =$ _____

_ _ _ _ _ _ _ _ _ _ _ _ _ _ _

675 ml, 675 ml

77

Isotonic dextrose solution can (elevate/lower) the CNS fluid pressure of a patient with a head injury.

_ _ _ _ _ _ _ _ _ _ _ _ _ _ _

elevate

78

Hypertonic solutions which can run faster than 3–4 ml/minute can

* _____.

_ _ _ _ _ _ _ _ _ _ _ _ _ _ _

damage the recipient's vein

79

Dextrose 5% in normal saline (0.9%) will (elevate/lower) the CNS fluid

pressure of a patient with a head injury. Explain why? _____

_____ .

_ _ _ _ _ _ _ _ _ _ _ _ _ _ _ _

lower, It is a hypertonic solution, pulling fluid from the brain.

80

What can happen when using large quantities of hypertonic solutions?

_____ .

_ _ _ _ _ _ _ _ _ _ _ _ _ _ _

dehydration and later shock

81

The nursing interventions for overhydration would be to:
1.
2.
3.
4.

_ _ _ _ _ _ _ _ _ _ _ _ _ _ _

1. Observe individuals for signs of overhydration, i.e., edema, dyspnea, coughing, and cyanosis.
2. Regulate the flow rate according to the individual's clinical condition.
3. Stop parenteral therapy when adverse signs are discovered.
4. Others from Table 21.

82

Individuals with a clinical picture of low serum proteins, low serum

sodium, and edema (increased body water) are prone to _____
by parenteral therapy.

_ _ _ _ _ _ _ _ _ _ _ _ _ _

overhydration

83

Explain how increased, difficult breathing, and coughing are clinical

signs of overhydration. _____

_____ .

_ _ _ _ _ _ _ _ _ _ _ _ _ _

With overhydration, fluid collects in the lungs causing pulmonary congestion. Difficult breathing and coughing result.

84

Explain how cyanosis can be an indication of overhydration. _____

_____ .

_ _ _ _ _ _ _ _ _ _ _ _ _

Increased fluid in the lungs causes poor lung ventilation. Therefore,
tissues are deprived of O_2.

85

The nurse should be constantly monitoring intravenous fluids. At the
same time the nurse notes the clinical signs and symptoms of electrolyte
imbalance. Match the clinical signs and symptoms with their electrolyte
imbalance.

_____ 1. Nausea, abdominal cramps,   a. sodium retention
          tachycardia, followed later by   b. potassium deficit
          bradycardia   c. potassium excess

_____ 2. Edema

_____ 3. Dizziness, muscular weakness, and
          arrhythmia

_ _ _ _ _ _ _ _ _ _ _ _ _

1. c
2. a
3. b

86

When a patient is receiving large quantities of sodium solution, the

nurse should observe for _____ caused by *_____.

When patients receive excessive parenteral therapy without electrolyte
supplements, (hypokalemia/hyperkalemia) could occur. Symptoms of

hypokalemia are _____ , _____ , and

_____ .

– – – – – – – – – – – – – – –

edema, sodium retention, hypokalemia; dizziness, arrhythmia, and
muscular weakness

87

The nurse checks the vital signs for adverse reaction. Arrhythmia could

indicate _____ .

Tachycardia followed later with bradycardia could indicate _____

_____ .

Tachycardia with a fall in blood pressure could indicate _____

_____ .

– – – – – – – – – – – – – – –

hypokalemia, hyperkalemia, hypernatremia

88

Your patient is receiving 1000 ml 5% dextrose in water with 40 mEq/L
of KCl. You check the Abbott drip counter and find the drops to be
90 per minute. What would your action be?
1.
2.
3.

– – – – – – – – – – – – – – –

1. Slow the fluids down to 45 gtts per minute.
2. Check the urinary output.
3. Check for symptoms of potassium intoxication.

Review

Q. 1
Why is it advisable to avoid subcutaneous infiltration during intravenous

therapy? _____ .

— — — — — — — — — — — — — —

To prevent damage to the surrounding tissues

_____

Q. 2
Intravenous needles left in site for over 48 hours can cause *_____

_____ .

— — — — — — — — — — — — — —

an inflammatory reaction or thrombophlebitis

_____

Q. 3
Parenteral fluids which are cold (from the refrigerator) when being ad-

ministered can cause _____ .

— — — — — — — — — — — — — —

vasoconstriction

_____

Q. 4
Hydrating solutions help in restoring *_____ .

— — — — — — — — — — — — — —

urinary output

_____

Q. 5
The use of large quantities of hydrating solutions in individuals with

poor circulation and chronic renal dysfunction could cause _____

_____ .

— — — — — — — — — — — — — —

overhydration

_____

Q. 6
Symptoms of lung congestion because of overhydration are:
1.
2.
3.
4.

_ _ _ _ _ _ _ _ _ _ _ _ _ _ _ _

1. edema
2. dyspnea
3. cyanosis
4. coughing

Q. 7
Amino acid solutions are often administered in cases of _____

_____ and _____ .

_ _ _ _ _ _ _ _ _ _ _ _ _ _ _ _

hypoproteinema and malnutrition

Q. 8
Adverse reaction to amino acid solutions are:
1.
2.
3.

_ _ _ _ _ _ _ _ _ _ _ _ _ _ _

1. nausea
2. feeling of warmth
3. flushing

Q. 9

Isotonic dextrose solutions (5% dextrose in water) administered to an individual with a head injury could cause the CNS fluid (spinal fluid) pressure to be (elevated/lowered).

What would happen to the spinal fluid pressure if a hypertonic solution were used? _____ . Why? _____

_____ .

_ _ _ _ _ _ _ _ _ _ _ _ _ _ _ _

elevated, lowered, An increase of osmolality due to the hypertonic solution will draw fluid from the CNS area.

Q. 10

Administering potassium to a patient having poor renal function could

cause potassium _____ .

_ _ _ _ _ _ _ _ _ _ _ _ _ _ _ _

intoxication or excess

Q. 11

Symptoms of sodium retention are _____ , _____ ,

and _____ .

_ _ _ _ _ _ _ _ _ _ _ _ _ _ _ _

edema, tachycardia, and a fall in blood pressure

Q. 12

Give at least 3 symptoms for hypokalemia and hyperkalemia as related to parenteral therapy.

Hypokalemia	Hyperkalemia
1. _____	1. _____
2. _____	2. _____
3. _____	3. _____

- - - - - - - - - - - - - - -

Hypokalemia	Hyperkalemia
1. arrhythmia	1. nausea
2. dizziness	2. abdominal cramps
3. muscular weakness	3. tachycardia—later bradycardia

SUMMARY QUESTIONS

The questions in this summary will test your knowledge of the material found in Chapter 4. If the answer is unknown, refer to the section in the chapter for further understanding and clarification.

1. What are the three basic objectives for parenteral therapy?
2. In giving parenteral fluid and electrolyte therapy, good kidney function is extremely important. Renal dysfunction could result in electrolyte (retention/excretion).
3. What are the four main classifications of parenteral solutions? Explain the rationale for using each group of solutions and give examples.
4. The solution resembling the electrolyte structure of plasma is

   _____ . Explain the effectiveness of blood in parenteral therapy.
5. List at least 3 conditions in which selected multiple electrolyte solutions are helpful in treatment.

   a. _____ b. _____ c. _____
6. What is another name for subcutaneous parenteral therapy? What type(s) of solution(s) is/are frequently used?

7. Define the terms hematocrit and B.U.N. A high hematocrit reading

would be an indication of _____ and a high B.U.N.

would be an indication of _____ .

8. What solutions would be effective in correcting metabolic acidosis, metabolic alkalosis, respiratory acidosis, and respiratory alkalosis?

9. Give the approximate rates of flow for administering intravenous fluid for the following:

　a. Maintenance therapy
　b. Replacement therapy
　c. Hydration therapy

10. Nurses frequently assume responsibility for parenteral therapy which has been ordered by the physician. Explain the nursing intervention and rationale for the following clinical considerations pertaining to intravenous therapy.

　a. Injection site
　b. Intravenous needles
　c. Cold and warm intravenous solutions
　d. Flow rate for:
　　1. Hydrating solutions
　　2. Amino acid solutions
　　3. Dextrose solutions
　　4. Hypertonic solutions
　e. Overhydration
　f. Intake and output
　g. Electrolyte imbalance
　h. Vital signs

References

Abbott Laboratories, *Fluid and Electrolytes*, North Chicago, Ill., 1968, pp. 25–30.

Baxter Laboratories, Inc., *Fluid Therapy*, Morton Grove, Ill., 1962, pp. 1–6.

"Blood Volume," *Pitoclinic*, XI (March, 1964), pp. 5–10 (Ames Co.).

Brunner, Lillian Sholtis, *et al.*, *Medical-Surgical Nursing*, Philadelphia, Pa.: J. B. Lippincott Co., 1964, pp. 169–171.

Burgess, Richard E., "Fluid and Electrolytes," *American Journal of Nursing*, LXV (October, 1965), pp. 93–94.

Dutcher, Isabel E. and Sandra B. Fielo, *Water and Electrolytes*, New York: The Macmillan Co., 1967, pp. 146–164.

Kleiner, Israel S. and James M. Orten, *Biochemistry*, 7th ed., St. Louis, Mo.: C. V. Mosby Co., 1966, pp. 658–663.

Levenstein, Brabara P., "Intravenous Therapy: A Nursing Specialty," *The Nursing Clinics of North America*, I (June, 1966), pp. 259–265.

Metheney, Norma M. and William D. Snively, *Nurses' Handbook of Fluid Balance*, Philadelphia, Pa.: J. B. Lippincott Co., 1967, pp. 116–139.

Montag, Midred and Ruth Swenson, *Fundamentals in Nursing Care*, 3rd ed., Philadelphia, Pa.: W. B. Saunders Co., 1959, pp. 297–314.

Shafer, Kathleen N., *et al.*, *Medical-Surgical Nursing*, 3rd ed., St. Louis, Mo.: The C. V. Mosby Co., 1964, pp. 64–65.

Smith, Dorothy W. and Claudia D. Gips, *Care of the Adult Patient*, Philadelphia, Pa.: J. B. Lippincott Co., 1963, pp. 142–146.

Statland, Harry, *Fluid and Electrolytes in Practice*, 3rd ed., Philadelphia, Pa.: J. B. Lippincott Co., 1963, pp. 155–164.

Chapter 5

Clinical Conditions of Fluid and Electrolyte Imbalance

BEHAVIORAL OBJECTIVES

Upon completion of this chapter, the student will be able to:

Explain the physiological factors leading to dehydration, water intoxication, edema, and shock.

State various clinical situations associated with

    (a) dehydration

    (b) water intoxication

    (c) edema

    (d) shock

And explain the clinical management needed to alleviate the four conditions.

Explain and observe clinically the signs and symptoms of those four conditions. (You may need some assistance in recognizing all these symptoms.)

Name at least two selected diuretics responsible for specific electrolyte deficits.

Differentiate between the four types of shock and be able to identify the type of shock in a clinical situation.

Explain how clinical management will differ according to the type of shock.

INTRODUCTION

Many of the disease entities are inclined to have some sort of fluid and electrolyte imbalance. Much of the imbalance is the result of certain clinical conditions, i.e, dehydration, water intoxication, edema, and shock. Each of these clinical conditions is discussed in this chapter, including the physiological factors involved, clinical considerations, and clinical management. Several of the clinical conditions have examples which add to the clarification and understanding of the fluid and electrolyte imbalance present and clinical management needed.

244

Again, it must be emphasized that the physician computes and orders fluid replacements. However, the nurse should understand these clinical conditions and reasons for their occurrence, and should be cognizant of physical changes that can occur before and during clinical management.

DEHYDRATION

1

Dehydration means the loss of water from the body. The water is lost mainly from the extracellular fluid compartment. Severe water loss from the body can cause a condition known as _____ .

— — — — — — — — — — — — — — —

dehydration

2

The fluid loss leads to a drop in blood volume and an increase in serum sodium. This is known as "hypertonic dehydration."

With the fluid loss, there is actually a decrease in the total amount of sodium remaining in the body, but the serum-sodium level will be elevated since the water has been lost in excess of sodium.

Hypertonic dehydration occurs when the serum-sodium level is (elevated/decreased) with the fluid loss.

— — — — — — — — — — — — — — —

elevated

3

When water has been lost in excess of sodium, the serum-sodium level would be _____ even though the total amount of sodium remaining in the body would be _____ .

— — — — — — — — — — — — — — —

elevated decreased

4

The plasma osmolality increases with the retained sodium, causing water to be drawn from the cells. With the elevation of serum sodium, the extracellular fluid becomes (hypertonic/hypotonic), resulting in a(an) (increase/decrease) in plasma osmolality which will cause a withdrawal

of fluid from the _____.

_ _ _ _ _ _ _ _ _ _ _ _ _ _ _

hypertonic, increase, cells or intracellular compartment

5

The hypertonic extracellular fluid will cause:

(   ) a. intracellular dehydration
(   ) b. intracellular hydration

Explain. _____

_____.

_ _ _ _ _ _ _ _ _ _ _ _ _ _ _

a. X
b. –
The hypertonic extracellular fluid will pull intracellular fluid from the cells by osmosis.

6

Potassium, magnesium, phosphates, and some protein are lost with the intracellular water.

Increased plasma osmolality results in intracellular fluid (loss/excess).

With intracellular fluid loss, the following electrolytes are also lost:

( ) a. potassium
( ) b. sodium
( ) c. magnesium
( ) d. chloride
( ) e. phosphate

Why? _____ .

– – – – – – – – – – – – – – – –

loss
a. X
b. –
c. X
d. –
e. X

These electrolytes are mainly found in the cells.

Table 22 describes the degrees of dehydration in relation to the percent of body weight loss, symptoms, and body water deficit by liter for a man weighing 150 pounds. Study this table carefully; be able to name the degrees of dehydration, their symptoms, the percentage of body weight loss, and an estimation of body fluid loss in liters. Hopefully, you will be able to recognize and identify degrees of dehydration which can occur to your patients during your clinical experience. Refer back to this table as you find it necessary.

TABLE 22 Degrees of Dehydration

Degrees of Dehydration	Percent of Body Weight Loss	Symptoms	Body Water Deficit by Liter
Mild dehydration	2%	1. Thirst	1–2 liters
Marked dehydration	5%	1. Marked thirst 2. Dry mucous membranes 3. Dryness and wrinkling of skin 4. Acid-base equilibrium towards greater acidity 5. Tachycardia as blood volume drops 6. Temperature – low grade elevation, e.g., 99°F. 7. Urine volume small, highly concentrated. Specific gravity 1.035 and up	3–5 liters
Severe dehydration	8%	1. Same symptoms as marked dehydration, plus: 2. Skin becomes flushed 3. Mental changes, e.g., restlessness, disorientation, delirium 4. Change of personality, e.g., irritability	5–10 liters
Fatal dehydration	22–30% total body water loss prove fatal	1. Coma 2. Death	

7

If water loss reaches 8% of the body weight, then severe dehydration results. Life will not continue with a water loss of _____% of body
weight.

The four degrees of dehydration are:
1.
2.
3.
4.

– – – – – – – – – – – – – – –

22–30%
1. mild
2. marked
3. severe
4. fatal

8

The percent of body weight loss is a guide for replacement fluid therapy.

What percent loss is associated with mild dehydration? _____.

The symptom for mild dehydration is _____.

– – – – – – – – – – – – – – –

2%, thirst

9

Symptoms for marked dehydration would be:

( ) a. Disorientation
( ) b. Dry mucous membranes
( ) c. Dryness and wrinkling of skin
( ) d. Irritability
( ) e. Marked thirst
( ) f. Delirium
( ) g. Tachycardia as blood volume drops
( ) h. Body temperature 97°
( ) i. Urine volume decreased
( ) j. Specific gravity of urine of 1.010
( ) k. Urine highly concentrated

What percent loss is associated with marked dehydration?_____.

— — — — — — — — — — — — — — — —

a. –   g. X
b. X   h. –
c. X   i. X
d. –   j. –
e. X   k. X
f. –
5%

10

Symptoms for severe dehydration would be:

( ) a. Bradycardia
( ) b. Tachycardia as blood volume drops
( ) c. Temperature 99.6°F
( ) d. Urine volume elevated
( ) e. Specific gravity of urine of 1.035 and higher
( ) f. Skin flushed
( ) g. Irritability
( ) h. Restlessness, disorientation
( ) i. Specific gravity of urine lower than 1.020
( ) j. Marked thirst

What percent loss is associated with severe dehydration? _____ .

----

a. —   f. X
b. X   g. X
c. X   h. X
d. —   i. —
e. X   j. X
8%

11

The percent of body weight loss is a guide for *_____
therapy.

----

replacement fluid

12

For the average body weight of 150 pounds, the amount of body water

loss for mild dehydration is _____ , for marked dehydration

is _____ , and for severe dehydration is _____ .

1–2 liters, 3–5 liters, 5–10 liters

Clinical Considerations

13

In early dehydration, the fluid loss is derived in equal quantities from both the extracellular and intracellular fluid spaces.

With chronic dehydration, which fluid loss becomes predominant, the extracellular or the intracellular? _____ .

― ― ― ― ― ― ― ― ― ― ― ― ― ―

intracellular

14

As a rule, hypernatremia results from water depletion. The amount of water loss is in excess to the amount of _____ loss.

If the urinary excretion of potassium is decreased, then (hypokalemia/ hyperkalemia) will result.

― ― ― ― ― ― ― ― ― ― ― ― ― ―

sodium, hyperkalemia

15

The hematocrit and the plasma proteins are elevated as a result of hemo-concentration (increased red blood cells and a decrease of fluid).

Hemoconcentration occurs with dehydration because * _____

_____ .

The hematocrit and the plasma proteins are (elevated/decreased) in marked dehydration.

― ― ― ― ― ― ― ― ― ― ― ― ― ―

there is an increase of red blood cells and a decrease of fluid.
elevated

16

In dehydration, the urine soon becomes concentrated with a specific gravity of 1.035 or higher. The urinary output is (increased/decreased).

Hypernatremia frequently results from *_____.

_ _ _ _ _ _ _ _ _ _ _ _ _ _ _

decreased, water depletion

17

Mild dehydration can result from fever, sweating, or insufficient water intake. Other causes of dehydration are:
    Diarrhea
    Vomiting, persistent
    Metabolic acidosis
    Hypovolemia due to:
        Hemorrhage
        Shock
        Burns

Hypovolemia may be a new term to you. Hypo would indicate _____

_____. Volemia comes from the Latin word, volumen, meaning volume.

    Actually hypovolemia is a diminished volume of circulating blood or intravascular fluid. Frequently, it is referred to as low blood volume.

_ _ _ _ _ _ _ _ _ _ _ _ _ _ _

low, less, deficit, or diminished

18

Some causes for mild dehydration are _____ , _____ ,

and _____ .

Other causes of dehydration are:

( ) a. Diarrhea
( ) b. Persistent vomiting
( ) c. Excessive water intake
( ) d. Hemorrhage
( ) e. Congestive heart failure
( ) f. Metabolic acidosis
( ) g. Increased caloric intake
( ) h. Burns

_ _ _ _ _ _ _ _ _ _ _ _ _ _ _

fever, sweating, and insufficient water intake

a. X   e. –
b. X   f. X
c. –   g. –
d. X   h. X

Clinical Management

In replacing body water loss, the total fluid deficit is estimated according to the percent of body weight loss. The physician computes the
fluid replacement for his patient. The following is only an example.
Many physicians use this method for replacement of fluid loss.

Clinical Example

Mr. Smith, who was admitted to the hospital, had a weight loss of 10
pounds due to dehydration. His weight had originally been 154 pounds,
or 70 Kg (kilograms). To determine the percent of body weight loss, one
would divide the weight loss by the original weight: therefore,
$10 \div 154 = .06$ or 6%. To determine the total fluid loss, one would
multiply the percent of body weight loss by Kg of body weight: therefore, $.06 \times 70$ Kg = 4.2 Liters.

19

Clinically, he would be considered to have:
( ) a. Mild dehydration
( ) b. Marked dehydration
( ) c. Severe dehydration

— — — — — — — — — — — — — — —

a. —
b. X
c. —

20

To determine the percent of body weight loss, one would *_____

_____.

To determine the total fluid loss, one would _____

_____.

— — — — — — — — — — — — — —

divide the weight loss by the original weight
multiply the percent of body weight loss by kilograms of body weight

21

One-third of body water deficit is from ECF (extracellular fluid) and
two-thirds of body water deficit is from ICF (intracellular fluid). To
determine replacement therapy, you would multiply:

$1/3 \times 4.2 \, L = 1.4 \, L$

Replacement fluid needed for ECF would be _____ ml.

$2/3 \times 4.2 \, L = 2.8 \, L$

Replacement fluid needed for ICF would be _____ ml.

2.5 L to replace the current day's losses.

The total fluid replacement for the first day would be _____.

— — — — — — — — — — — — — —

1400 ml, 2800 ml, 6.7 L or 6700 ml

22

One-third of the water deficit would be from the *_____

_____ and two-thirds of the water deficit from the

*_____ .

_ _ _ _ _ _ _ _ _ _ _ _ _ _ _

extracellular fluid, intracellular (cellular) fluid

23

The sodium deficit would be the amount contained in the extracellular

fluid loss of 1.4 L. Can you explain why the deficit? _____

_____

_____ .

   The potassium deficit would be the amount contained in the intracellular fluid loss of 2.8 L. Can you explain why the potassium deficit?

_____

_____

_____ .

_ _ _ _ _ _ _ _ _ _ _ _ _ _ _

Sodium is the main cation of ECF, so with ECF loss, Na would accompany it. However, the serum-sodium level may be elevated if the fluid
loss is greater than Na loss.
Potassium is the main cation for ICF, so with ICF loss, K would accompany it. However, the serum-potassium level may be elevated because
the K leaves the cells and accumulates in the ECF.

Suggested Solutions for Administration

First: Replace fluid volume and reduce osmolality.

24

In case of a low plasma CO_2 combining power—metabolic acidosis—the solutions should contain excess bicarbonate rather than chloride. Can

you explain why? _____

_____ .

_ _ _ _ _ _ _ _ _ _ _ _ _ _ _ _

A low plasma CO_2 indicates a lack of the HCO_3 ion which is needed to neutralize the body's acidotic state. Administering Cl would combine with H ion and increase acidosis.

25

The following suggested solution replacement would be needed in correcting dehydration.

   1500 ml of lactated Ringer's—to replace ECF losses.
   500 ml of M/6 sodium lactate to make up sodium deficit and help correct metabolic acidosis. This solution will increase plasma CO_2.
   4700 ml of 5% dextrose water—to replace water deficit and increase urinary output. Potassium chloride 120 mEq is added to 3 L to replace potassium loss.

When the potassium is restored in the cells, would the intracellular fluid

be increased or decreased? _____ . Why do you think

this might happen? _____

_____ .

   When potassium is being administered in the form of KCl or potassium chloride, explain your concern about the patient's urinary output.

_____ .

_ _ _ _ _ _ _ _ _ _ _ _ _ _ _ _

increased, Fluid would flow into the cells as potassium returns to the cells. Cellular function would be restored. It should be sufficient, at least 250 ml every 8 hours. Poor urinary output leads to potassium excess.

26

In correcting dehydration, the first steps would be to *_____

_____and *_____.

– – – – – – – – – – – – – – – –

replace fluid volume and reduce osmolality

27

Lactated Ringer's solution is helpful in treating dehydration because:
( ) a. It resembles the electrolytic structure of normal blood serum.
( ) b. It will replace the extracellular fluid volume.
( ) c. It will replace the all electrolyte loss.
( ) d. It will replace potassium loss.

M/6 sodium lactate is helpful in treating dehydration because:
( ) a. It replaces the sodium loss.
( ) b. It will aid in decreasing the CO_2 combining power of the plasma.
( ) c. It will aid in increasing the CO_2 combining power of the plasma.
( ) d. It is helpful in the correction of metabolic acidosis.
( ) e. It is helpful in the correction of metabolic alkalosis.

Dextrose 5% in water is helpful in treating dehydration because:
( ) a. It will replace water deficit.
( ) b. It will aid in increasing urine output.
( ) c. It will aid in decreasing urine output.
( ) d. It will replace the sodium deficit.

– – – – – – – – – – – – – – –

a. X	a. X	a. X
b. X	b. –	b. X
c. –	c. X	c. –
d. –	d. X	d. –
	e. –	

28

Mild dehydration is frequently treated with dextrose, water, and small amounts of electrolytes.

Dextrose 5% in water is frequently given first, followed by a solution of low electrolyte content with 5% dextrose. An example of this solution could be _____. (These solutions could be given in reverse according to patient's condition and physician's choice.)

    When administering 5% dextrose in water, dextrose is metabolized quickly, leaving _____ to replace fluid deficit.

— — — — — — — — — — — — — — —

lactated Ringer's or Ringer's solution, water

Review—Dehydration

Q. 1

Hypertonic dehydration results from a decrease in * _____

_____ and an increase in _____ .

— — — — — — — — — — — — — — —

blood volume of body fluid, serum sodium

Q. 2

Why does plasma osmolality increase with dehydration? _____

_____ . What results from this increase? _____

_____ .

— — — — — — — — — — — — — — —

Because of retained Na. Water is drawn from the cells due to the process of osmosis.

Q. 3

Increase in plasma osmolality can result in *_____

_____ .

_ _ _ _ _ _ _ _ _ _ _ _ _ _ _

cellular dehydration, or loss of water from the cells, or similar response

Q. 4

The main symptom for mild dehydration is _____ .

_ _ _ _ _ _ _ _ _ _ _ _ _ _ _

thirst

Q. 5

Give at least five symptoms of severe dehydration.
1.
2.
3.
4.
5.

_ _ _ _ _ _ _ _ _ _ _ _ _ _ _

1. marked thirst
2. decreased urine volume
3. tachycardia
4. flushed skin
5. disorientation, irritability, or any others

Q. 6

Hypernatremia frequently results from *_____ .

_ _ _ _ _ _ _ _ _ _ _ _ _ _ _

water depletion

Q. 7

In dehydration, why are the hematocrit and the plasma proteins elevated?

_____

_____ .

– – – – – – – – – – – – – – – –

Because of hemoconcentration—an increase of red blood cells and a decrease of body fluid.

Q. 8

Some causes for mild dehydration would include:

1.

2.

3.

– – – – – – – – – – – – – – –

1. fever
2. sweating
3. insufficient water intake

Q. 9

In correcting dehydration, the considerations would be to replace

* _____ and to reduce * _____ .

– – – – – – – – – – – – – – –

fluid volume, plasma osmolality

Q. 10

Lactated Ringer's solution is helpful in treating dehydration for it resembles * _____ and it will replace * _____

_____ .

– – – – – – – – – – – – – –

the electrolyte structure of plasma, extracellular fluid volume

Q. 11

M/6 sodium lactate is helpful in treating dehydration for it replaces

* _____ and will _____ the CO_2 combining power of the plasma.

_ _ _ _ _ _ _ _ _ _ _ _ _ _ _

sodium loss, increase

Q. 12

Dextrose 5% in water is helpful in treating dehydration for it will:
1.
2.

_ _ _ _ _ _ _ _ _ _ _ _ _ _ _

1. replace water deficit
2. increase the urinary output

WATER INTOXICATION

29

Water intoxication occurs as a result of an excess water intake and a lack of sodium intake. Also associated with this condition is hypotonicity, or a rapid decrease in serum osmolality.

With water intoxication, would the sodium intake be greater or lesser

than the water intake? _____ .

As the result of this imbalance, explain what would happen to the

serum osmolality? _____ .

_ _ _ _ _ _ _ _ _ _ _ _ _ _ _

lesser, It would decrease, or hypotonicity

30

With dehydration, the water loss could be greater than the sodium loss, so with this, what would be the serum osmolality? _____

_____ .

_ _ _ _ _ _ _ _ _ _ _ _ _ _ _

It would be increased, or hypertonicity.

31

Water intoxication is not the same as edema. Edema is the accumulation of fluid in the interstitial spaces. With water intoxication, the excess fluid first enters the extracellular space, lowering the osmotic pressure. Then, due to osmosis, water moves from the extracellular fluid into the cells, causing the cells to swell.

In water intoxication, the water first enters the *_____

_____ .

Why would the water then leave the extracellular space and move into the cells? _____

_____ .

_ _ _ _ _ _ _ _ _ _ _ _ _ _ _

extracellular space, Due to osmosis, the cells have the greatest concentration, drawing water into the cells.

32

Water intoxication (is/is not) the same as edema. Edema is an extracellular accumulation of fluid, whereas with water intoxication, the end result would be an (extracellular/intracellular) accumulation of fluid. This causes the cells to _____ .

_ _ _ _ _ _ _ _ _ _ _ _ _ _ _

is not, intracellular, swell

Clinical Considerations

33

It is difficult for an individual to drink himself into water intoxication unless the renal mechanisms for elimination fail.

If excessive water has been given and the kidneys are not functioning

properly, what would likely occur? _____

_____ .

_ _ _ _ _ _ _ _ _ _ _ _ _ _ _ _ _ _

water retention, or water intoxication

34

The most common occurrence of water intoxication is seen in post-operative patients when fluids orally and intravenously have been forced without compensatory amounts of salt. The amount of water exceeds that which the kidneys can excrete.

Also, following surgery an overproduction of the antidiuretic hormone (ADH) frequently occurs due to trauma and anesthesia. Because of the overproduction of ADH, the water excretion will (increase/decrease), causing the urine volume to (rise/drop).

_ _ _ _ _ _ _ _ _ _ _ _ _ _ _ _

decrease, drop

35

When does water intoxication most commonly occur? _____

_____ . Why? _____

_____

_____ .

_ _ _ _ _ _ _ _ _ _ _ _ _ _ _ _

With postoperative patients. Frequently, when forced fluids have been given without compensatory amounts of salt, the kidneys are unable to excrete the excess water.

36

If excess ADH is the cause, the specific gravity of urine is high and the urinary output in 24 hours may be as low as 500–700 ml.

If drinking excess water (as in psychogenic polydipsia) or forced fluids postoperatively is the cause, the specific gravity will be low. What two factors might cause an overproduction of ADH (antidiuretic hormone)

following surgery? _____ and _____ .

In case of excess ADH, will specific gravity of urine be increased or

decreased? _____ .

Will urine volume be increased or decreased? _____ .

— — — — — — — — — — — — — — —

trauma and anesthesia, increased, decreased

37

If the circulation through the kidneys is impaired, this will cause a reten-

tion of water which can lead to *_____ .
Impairment of circulation through the kidneys can be due to arterio-sclerosis. If the kidneys do not receive sufficient blood circulation, kidney dysfunction can result.

— — — — — — — — — — — — — — —

water intoxication

38

Causes of water intoxication are:
1. Excessive water intake postoperatively
2. Kidney dysfunction
3. Overproduction of ADH
4. Psychogenic polydipsia
Can you think of another cause for water intoxication? _____

_____ .

— — — — — — — — — — — — — — —

Impairment of circulation through the kidneys.

266 Clinical Conditions of Fluid and Electrolyte Imbalance

39
Name the five causes for water intoxication.
1.
2.
3.
4.
5.

----

1. Kidney dysfunction
2. Forced fluids
3. Overproduction of ADH
4. Psychogenic polydipsia
5. Impairment of circulation

Symptoms of Water Intoxication

A. Early Symptoms
   Headache, nausea and vomiting, and excessive perspiration
B. Progressive Symptoms
   CNS (Central Nervous System). Most prominent symptoms include:
      Behavior changes
      Incoordination
      Drowsiness
      Aphasia
      Confusion
C. Later Symptoms
   CNS symptoms include:
      Neuroexcitability—muscle twitching
      Delirium
      Convulsions—then coma
   Skin—warm, moist, and flushed

40
What are the early symptoms of water intoxication?
1.
2.
3.

— — — — — — — — — — — — — — —

1. headache
2. nausea and vomiting
3. excessive perspiration

41
Central Nervous System symptoms are (least/most) prominent with
water intoxication.

— — — — — — — — — — — — — — —

most

42
Refer to Symptoms of Water Intoxication if necessary. Examples of progressive and later CNS symptoms include:
1.
2.
3.
4.
5.

— — — — — — — — — — — — — — —

1. behavior changes
2. incoordination
3. drowsiness
4. aphasia
5. confusion
6. muscle twitching or neuroexcitability
7. delirium
8. convulsions, coma

43
The skin in later stages of water intoxication is *_____

_____ .

‒ ‒ ‒ ‒ ‒ ‒ ‒ ‒ ‒ ‒ ‒ ‒ ‒ ‒ ‒

warm, moist, flushed

44
Name at least three progressive CNS symptoms of water intoxication:
1.
2.
3.

‒ ‒ ‒ ‒ ‒ ‒ ‒ ‒ ‒ ‒ ‒ ‒ ‒ ‒ ‒

1. behavior changes
2. incoordination
3. drowsiness
4. aphasia
5. confusion

Clinical Management

Overall Objective: To reduce excess water in the body.
  There are two ways to reduce this water:
      1. Reduce water intake
      2. Promote water excretion

45

In less severe cases of water intoxication, water restriction may be sufficient, or an extracellular replacement solution such as lactated Ringer's may be given to increase the osmolality of the extracellular fluid.

What is the overall objective in the clinical management of water intoxication? _____ .

Name two ways in which this objective can be accomplished.

1.

2.

_ _ _ _ _ _ _ _ _ _ _ _ _ _ _ _

To reduce excess water in the body.

1. reduce water intake

2. promote water excretion

46

Concentrated saline may be given in more severe cases of water intoxication to raise extracellular electrolyte concentration in hope of drawing water out of the (intracellular space/interstitial space) and (increasing/decreasing) urinary output.

_ _ _ _ _ _ _ _ _ _ _ _ _ _ _ _

intracellular space, increasing

47

However, administration of additional salt to an individual who already has too much water could result in expansion of the interstitial fluid and blood volume, and the development of (water intoxication/edema).

An osmotic diuretic, e.g., mannitol, induces diuresis and a loss of retained fluid.

_ _ _ _ _ _ _ _ _ _ _ _ _ _

edema

48

From Frames 45, 46, and 47, can you give three methods for promoting water excretion?

1.

2.

3.

_ _ _ _ _ _ _ _ _ _ _ _ _ _

1. Extracellular replacement solution, e.g., lactated Ringer's
2. Concentrated saline solution
3. Osmotic diuretics, e.g., mannitol

49

For less severe cases of water intoxication, what would the clinical management be? _____ and/or

_____ .

For more severe cases of water intoxication, what would the clinical management be? _____ and/or

_____ .

_ _ _ _ _ _ _ _ _ _ _ _ _ _

water restriction and/or extracellular replacement solution
concentrated saline solution and/or osmotic diuretics

50

An osmotic diuretic induces * _____

_____ .

_ _ _ _ _ _ _ _ _ _ _ _ _ _

diuresis and a loss of retained fluid

Review—Water Intoxication

Q. 1

Water intoxication occurs as a result of * _____

and * _____ .

_ _ _ _ _ _ _ _ _ _ _ _ _ _ _

an excess water intake and a lack of sodium intake

Q. 2

In water intoxication, the serum osmolality would be (increased/

decreased). What would be the tonicity of the serum? _____ .

_ _ _ _ _ _ _ _ _ _ _ _ _ _ _

decreased, hypotonicity

Q. 3

With edema, fluid accumulates in the interstitial spaces, but with water
intoxication, fluid first accumulates in the intravascular space and later

in the _____ .

_ _ _ _ _ _ _ _ _ _ _ _ _ _ _

cells or intracellular space.

Q. 4

With more severe water intoxication, what happens to the cells?

_____ .

_ _ _ _ _ _ _ _ _ _ _ _ _ _ _

They swell.

Q. 5

Water intoxication frequently occurs in the postoperative patients who

are on forced liquids without * _____ .

_ _ _ _ _ _ _ _ _ _ _ _ _ _ _

sufficient salt

Q. 6

Also, water intoxication can occur when there is an overproduction of the antidiuretic hormone (ADH), which can occur following surgery when there is _____ and _____ .

_____

trauma and anesthesia

Q. 7

Give the five causes for water intoxication:

1.
2.
3.
4.
5.

_____

1. kidney dysfunction
2. forced fluids, postoperative
3. overproduction of ADH
4. psychogenic polydipsia
5. impairment of circulation

Q. 8

Name three early symptoms of water intoxication: _____ ,

_____ , and _____ .

_____

headache, nausea, vomiting, and excessive perspiration

Q. 9

The CNS symptoms are (less/most) prominent with water intoxication.

_____

most

Q. 10

In clinical management, the ways for reducing water intoxication are to:
1.
2.

_ _ _ _ _ _ _ _ _ _ _ _ _ _ _

1. promote water excretion
2. reduce water intake

Q. 11

Name the three methods for promoting water excretion.
1.
2.
3.

_ _ _ _ _ _ _ _ _ _ _ _ _ _

1. extracellular replacement solution
2. concentrated saline solution
3. osmotic diuretics

EDEMA

51

Edema is the abnormal retention of fluid in interstitial spaces or in serous cavities.

Edema results from sodium retention in the body, causing a retention of water and an increase in the extracellular fluid volume.

Edema is the result of sodium excess, whereas water intoxication is

the result of *_____. With water intoxication

would there be excess or insufficient salt? _____.

_ _ _ _ _ _ _ _ _ _ _ _ _ _ _

excess water intake, insufficient

52

Edema is the abnormal retention of fluid in the *_____

_____ .

_ _ _ _ _ _ _ _ _ _ _ _ _ _ _

interstitial spaces or in serous cavities

53

Edema results from the excess of what ion? _____ .

_ _ _ _ _ _ _ _ _ _ _ _ _ _ _

sodium

Diagram 11 demonstrates the make-up of normal body fluid versus abnormal body fluid, such as with edema. As you recall from Chapter 1, 60% of the adult body weight is water; 40% of that is intracellular or cellular water and 20% is extracellular water. Of the extracellular fluid, 15% is interstitial fluid and 5% is intravascular fluid or plasma. Note that with edema there is an increase of fluid in the interstitial space, which is between tissues and cells. The intracellular fluid may be decreased in extreme cases. Refer to this diagram as needed.

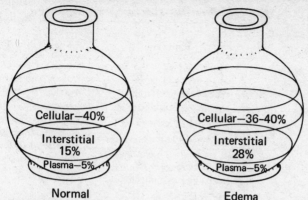

Normal
Fluid Percent of Body Weight

Edema
Fluid Percent of Body Weight

Diagram 11. Body fluid compartments and edema. (Adapted from Harry Statland, *Fluid and Electrolytes in Practice,* J. B. Lippincott Co., Philadelphia, Pa., p. 177.)

54

In edema, the greatest increase in volume occurs in which of the fluids?

_____.

– – – – – – – – – – – – – – –

interstitial fluid

55

The intracellular fluid may be (increased/decreased) in extreme cases.

– – – – – – – – – – – – – –

decreased

Table 23 gives five physiological factors that can cause edema. With each physiological factor, the rationale and examples of clinical situations in which edema will occur are given.

Study this table carefully; note whether there is an increase or decrease in the physiological factors which can serve as an indication of edema. Be able to explain how edema occurs in the various clinical situations. Refer to this table as you find it necessary.

TABLE 23 Physiological Factors Leading to Edema

Physiological Factors	Rationale	Clinical Situations
Plasma hydrostatic pressure in the capillaries	↑ Increased Blood dammed in the venous system will cause "back" pressure on the capillaries, thus raising the capillary pressure. An increase in the pressure of the blood in the capillaries will force more fluid into the tissue areas, thus producing edema.	1. Congestive heart failure with increase in venous pressure. 2. Venous obstruction leading to varicose veins. 3. Pressure on the veins because of swelling, constricting bandages, or casts.
Plasma colloid osmotic pressure	↓ Decreased The decrease in plasma colloid osmotic pressure results from diminished plasma protein concentration. A decrease in protein content will cause water to flow from the plasma into the tissue spaces, thus causing edema.	1. Malnutrition due to lack of protein in the diet. 2. Chronic diarrhea resulting in loss of protein. 3. Burns leading to loss of fluid containing protein through denuded skin. 4. Kidney disease, particularly nephrosis. 5. Loss of plasma proteins through the urine.
Capillary permeability	↑ Increased An increase in the permeability of the capillary membrane will allow the plasma proteins to leak out of the capillaries into the interstitial space more rapidly than the lymphatics can return them to the circulation. Increased capillary permeability is a predisposing factor to edema.	1. Bacterial inflammation causes increased porosity. 2. Allergic reactions. 3. Burns causing damage to the capillaries. 4. Acute kidney disease, e.g., nephritis.

TABLE 23 (continued)

Physiological Factors		Rationale	Clinical Situations
Sodium retention	↑ Increased	Kidneys regulate the level of sodium ions in the extracellular fluid. Kidney function will depend upon an adequate blood flow. Inadequate blood flow, presence of excess aldosterone, or diseased kidneys, are predisposing factors to edema since these cause sodium retention.	1. CHF, congestive heart failure, causing inadequate circulation of blood. 2. Renal failure—Inadequate circulation of blood through the kidneys. 3. Increased production of the adrenal cortical hormones —aldosterone, cortisone, and hydrocortisone—will cause retention of sodium. 4. Cirrhosis of the liver. The diseased liver cannot destroy excess production of aldosterone. 5. Trauma resulting from fractures, burns, surgery.
Lymphatic drainage	↓ Decreased	Blockage of the lymphatics will prevent the return of proteins to the circulation. The obstructed lymph flow is said to be high in protein content. With an inadequate return of proteins to the circulation, the plasma colloid osmotic pressure will be decreased, thus causing edema.	1. Lymphatic obstruction, e.g., cancer of the lymphatic system. 2. Surgical removal of lymph nodes. 3. Elephantiasis which is a parasitic invasion of the lymph channels, resulting in fibrous tissue growing in the nodes, obstructing the lymph flow. 4. Obesity because of inadequate supporting structures for the lymphatics in the lower extremities. The muscles are considered the supporting structures.

56

Place I for increased and D for decreased concerning the physiological factors as they occur in edema:

_____ a. Capillary permeability

_____ b. Sodium retention

_____ c. Lymphatic drainage

_____ d. Plasma colloid osmotic pressure

_____ e. Plasma hydrostatic pressure in the capillaries

_ _ _ _ _ _ _ _ _ _ _ _ _ _ _ _ _

a. I
b. I
c. D
d. D
e. I

57

Blood backed up in the venous system will increase the capillary pressure, forcing more fluid into *_____

_____.

   The clinical situations in which edema may occur as a result of an increase of plasma hydrostatic pressure are:

1.

2.

3.

_ _ _ _ _ _ _ _ _ _ _ _ _ _ _

the tissue areas (interstitial areas)
1. CHF, congestive heart failure
2. venous obstruction
3. pressure on the veins, e.g., casts or bandages

58

A decrease in plasma protein will result in a decrease in the plasma colloid osmotic pressure. This will cause water to flow from the plasma

into *_____.

Name at least three situations in which edema occurs as a result of a decrease in plasma colloid osmotic pressure:

1. _____
2. _____
3. _____

– – – – – – – – – – – – – – – – –

the tissue spaces (interstitial spaces)
1. malnutrition
2. chronic diarrhea
3. kidney diseases
4. burns
5. urinary excretion of protein

59

An increase in the capillary membrane permeability will permit plasma

proteins to escape from _____ causing more water to flow

into *_____.

Name at least three situations in which edema occurs as a result of increased capillary permeability:

1. _____
2. _____
3. _____

– – – – – – – – – – – – – –

capillaries, the tissue spaces (interstitial spaces)
1. bacterial inflammation
2. allergic reactions
3. acute kidney disease, e.g., nephritis
4. burns

60

The kidneys regulate the level of _____ ions in the extracellular
fluid. An inadequate blood flow, the presence of excess aldosterone, or
diseased kidneys will result in sodium (excretion/retention).

   Name at least three clinical situations which can cause sodium retention:
1.
2.
3.

— — — — — — — — — — — — — — —

sodium, retention
1. CHF, congestive heart failure
2. renal failure
3. adrenal cortical hormones, e.g., cortisone
4. cirrhosis of the liver
5. trauma

61

Obstruction of the lymph flow will prevent the return of proteins to

the circulation. The obstructed lymph flow is high in _____
content.

   A decrease in protein content in the plasma will cause the water to

flow from * _____ into * _____

_____.

   Name at least three clinical situations which cause a decrease in
lymphatic drainage:
1.
2.
3.

— — — — — — — — — — — — — — —

protein, the plasma or intravascular space into the tissue spaces
1. cancer of the lymphatic system
2. removal of the lymph nodes
3. obesity
4. elephantiasis

Review—Causes of Edema

This review should aid in reinforcing your understanding of the physiological factors leading to edema. Not all possible answers have been given, so after you have completed this review, check your answers against Table 23.

Place I for increased and D for decreased concerning the physiological factors as they occur in edema. Explain the rationale for each physiological factor and give an example of a clinical situation.

_____Hydrostatic pressure in the capillaries.

_____

_____

Example: _____

I. Blood jammed in the venous system will increase capillary pressure.
Example: CHF or varicose veins

_____Plasma colloid osmotic pressure.

_____

_____

Example: _____

D. Lack of protein will cause a decrease in the plasma colloid osmotic pressure, thus water will flow into the tissue. Example: burns, chronic diarrhea

_____Capillary permeability

_____

_____

Example: _____

I. An increased capillary membrane permeability will permit plasma protein to escape from capillaries. Example: burns, inflammation

_____

_____Sodium retention.

Example: _____

I. Inadequate blood flow, presence of excess aldosterone, or diseased
kidneys will cause sodium retention. Example: CHF, Cushing Syndrome

_____Lymphatic drainage.

Example: _____

D. Obstruction of the lymph flow will prevent the return of proteins
to the circulation. Example: removal of lymph nodes, elephantiasis

Clinical Considerations

62

The influence of gravity has an effect on the distribution of edema fluid
in the edematous individual. In a lying-down position, there is a more
equal distribution of edema, whereas in an upright position, the edema
would be more prevalent in the lower extremities.

The eyelids of a person with generalized edema would be swollen in
the morning, but by afternoon, with increased activity, the swelling
would be (more/less) marked.

_ _ _ _ _ _ _ _ _ _ _ _ _ _ _ _

less

63

In the edematous individual, the distribution of edema fluid is influenced

by _____.

_ _ _ _ _ _ _ _ _ _ _ _ _ _ _

gravity

64

With individuals who are up and about, the edema fluid frequently will
be found in the (ankles and feet/sacrum and buttocks). For those who

are bedridden, edema fluid will most likely be found at the *_____

_____.

_ _ _ _ _ _ _ _ _ _ _ _ _ _ _ _

ankles and feet
eyes or sacrum and buttocks, or more equally distributed

65

Edema of the lungs, often called pulmonary edema, can occur in
patients with limited cardiac or renal reserve. If the heart is not able to
beat adequately and the kidneys cannot excrete a sufficient amount of
urine, then the fluid will back up into the pulmonary circulatory system.

  When the hydrostatic pressure of the blood in the pulmonary capillaries rises to equal or exceed the plasma colloid osmotic pressure, then

the water will flow from plasma into the *_____
leading to pulmonary edema.

_ _ _ _ _ _ _ _ _ _ _ _ _ _ _

lung tissues

66

The individual with pulmonary edema (fluid throughout the lung tissue)
has moist râles and shortness of breath. The alveoli (air sacs) are filled
with fluid.

  What effect do you think fluid throughout the lung tissue would have

on ventilation? _____.

_ _ _ _ _ _ _ _ _ _ _ _ _ _ _

poor or inadequate ventilation

67

Fluid throughout the lung tissue is known as *_____.
  Individuals with poor cardiac output are prone to develop pulmonary

edema. Can you explain why? _____

_____.

– – – – – – – – – – – – – – –

pulmonary edema
Fluid (intravascular) "backs up" into the pulmonary circulation since
the heart output is insufficient.

68

Giving excessive intravenous infusions to an individual with pulmonary
edema would cause the blood volume to increase. What do you think

increased blood volume is called? _____.
  Intravenous infusions should be regulated so that the rate of flow is

not in excess to the urinary _____ .

– – – – – – – – – – – – – – –

hypervolemia, output

69

Giving excessive parenteral fluids to an individual in severe congestive
heart failure will cause an increase in pulmonary venous pressure.

  This will predispose the individual to *_____.

  A name for increased blood volume is _____.

– – – – – – – – – – – – – – –

pulmonary edema, hypervolemia

70

The tissues of an edematous person are said to be more vulnerable to injury resulting in tissue breakdown.

A bedfast individual with edema of the sacrum and buttocks is apt to

develop _____ due to * _____ .

- - - - - - - - - - - - - - -

decubiti (bedsores), tissue breakdown

71

Generalized edema is called anasarca. The following terms are given in describing edema fluid in the various body cavities:

    Peritoneal cavity: ascites
    Pleural cavity: hydrothorax
    Pericardial sac: hydropericardium

Anasarca means _____ .
Give the names of edema fluid in the following body cavities:

    Peritoneal cavity: _____

    Pleural cavity: _____

    Pericardial sac: _____

- - - - - - - - - - - - - - -

generalized edema, ascites, hydrothorax, hydropericardium

Clinical Management for Edema

72

With nephrotic edema or any other marked hypoproteinemic state, the administration of albumin, which is (hypotonic/hypertonic), will (raise/ decrease) the plasma colloid osmotic pressure, this would cause the

fluid to flow from the * _____ into the _____ .

- - - - - - - - - - - - - - -

hypertonic, raise, tissue space, plasma

Clinical Example
73

Mrs. Allen was admitted to the hospital in congestive heart failure. She had shortness of breath when walking up a flight of stairs and had swelling of the ankles. The physiological factors were:

1. Her heart was no longer able to fully overcome the resistance to the flow of blood by her constricted arteries and so the circulation was diminished. This increased the venous (hydrostatic) pressure.
2. An increase in the production of aldosterone and ADH (antidiuretic hormone) resulted in a sodium retention, which caused the volume of her extracellular fluid to (rise/decrease).
3. With an elevation in venous pressure and sodium retention, the kidneys were unable to remove the necessary amount of _____

   and _____ .
4. The result—(edema/water intoxication).

— — — — — — — — — — — — — — —

rise, water, sodium, edema

74

Since the cause of edema formation in this case is the retention of sodium, along with a large amount of water, the ultimate aim would be to reduce the salt and water intake.

   Water intake alone probably (will/will not) increase the edema.

   Excess salt intake with water probably (will/will not) increase the edema.

— — — — — — — — — — — — — — —

will not, will

The "Three D's" are frequently employed in the management of edema:
1. Diet
2. Digitalization
3. Diuretics

The clinical management of edema in Mrs. Allen's case incorporated the "Three D's."

75

Diet

Mrs. Allen was placed on a low-sodium diet and her fluid intake was limited to 1200 ml (300 ml below daily requirement).

By limiting salt and water intake this would
( ) a. increase edema
( ) b. prohibit further increase of edema
( ) c. decrease water intoxication

- - - - - - - - - - - - - - - -

a. —
b. X
c. —

76

Digitalization

Digitilization is the process of increasing the serum level of digitalis to achieve the desired physiological effect.

Digitalis preparations, which are under the classification of cardiotonics, will slow down the ventricular contractions and make them more forceful. Examples of digitalis preparations are: Digoxin, Digitoxin, Gitalin, Cedilanid, and Digitalis Leaf. Digoxin is frequently the choice cardiotonic for digitalizing individuals having poor cardiac output.

Digitalis is classified as a _____. This drug will slow

down the * _____ and make the heart beat

* _____ .

Mrs. Allen was digitalized with Digoxin and then placed on a daily maintenance dose of Digitoxin 0.2 mg. This would (increase/decrease) cardiac output.

Blood circulation would then be _____ .

The urinary output would be _____ .

It is important that you remember the toxic effects of digitalis preparations, which are pulse below 60, nausea, vomiting, and/or diarrhea.

- - - - - - - - - - - - - - - -

cardiotonic, ventricular contractions, more forceful, increase, improved or increased, increased

77
Diuretics
Drugs used to increase the secretion of urine.

Many of the diuretics will not only cause excretion of sodium and water, but also of other valuable electrolytes, e.g., potassium, frequently chloride, and in some cases bicarbonate.

What are diuretics? _____ .
Diuretics will cause an excretion of sodium and water and can cause

a loss of what other electrolytes? _____ , _____ ,

and _____ .

– – – – – – – – – – – – – – –

Drugs used to increase the secretion of urine.
potassium, chloride, and bicarbonate

78
Osmotic, mercurial, sulfamyl, and thiazide diuretics when used over a prolonged period of time can frequently cause a serum potassium deficit.

Aldosterone antagonists used as diuretics will cause a loss of sodium and chloride in the urine without a loss of potassium.

Aldosterone is an adrenal cortical hormone which will cause sodium

to be _____ and potassium to be _____ .
Aldosterone antagonists will cause what two ions to be excreted?

_____ and _____ .

Which ion will be retained? _____ .

– – – – – – – – – – – – – – –

retained, excreted, sodium and chloride, potassium

79
Frequently physicians will prescribe a thiazide diuretic and an aldosterone antagonist to prevent excessive loss of what ion? _____ .

– – – – – – – – – – – – – – –

potassium

80

Diuretics which can cause a serum potassium deficit when used frequently or in excess are:

1.
2.
3.
4.

— — — — — — — — — — — — — — —

1. osmotic
2. mercurial
3. sulfamyl
4. thiazide

Table 24 gives the classification, generic and trade names of selected diuretics. You are given only the most commonly used diuretics and those which do have an effect on the body's electrolyte balance. Not included are several other classifications and diuretics that are not as commonly used. The generic names of drugs reflect the chemical family to which the drug belongs. It is never changed and can be used in all countries. The trade names are given by their manufacturer.

You are not asked to memorize this table for it serves as a reference table, and you may refer to it as needed.

TABLE 24   Selected Diuretics

Generic Name	Trade Name
Mercurial Diuretics	
Chlormerodrin	Neohydrin
Meralluride	Mercuhydrin
Mercaptomerium sodium	Thiomerin sodium
Sulfamyl Diuretics	
Acetazolamide	Diamox
Thiazide Diuretics	
Chlorothiazide	Diuril
Hydrochlorothiazide	Hydrodiuril or Esidrix
Osmotic Diuretics	
Mannitol	Osmitrol
Aldosterone Antagonist	
Spironolactone	Aldactone
Potent Miscellaneous Diuretics	
Furosemide	Lasix
Ethacrynic Acid	Edecrin

81

Mrs. Allen received hydrochlorothiazide for her edema. While she is receiving this diuretic the nurse should observe for which specific

electrolyte deficit? _____ .

_ _ _ _ _ _ _ _ _ _ _ _ _ _ _

potassium

82

Give at least five symptoms of potassium deficit. (Refer to Table 5 if necessary.)

1.
2.
3.
4.
5.

— — — — — — — — — — — — —

1. anorexia
2. muscular weakness
3. arrhythmia
4. dizziness (hypotension)
5. malaise
6. abdominal distention
Others listed in Table 5.

83

Indicate which of the following foods would be high in potassium. (Refer to Table 14 if necessary.)

( ) a. Orange juice
( ) b. Vegetables
( ) c. Bananas
( ) d. Meats
( ) e. Nuts
( ) f. Bread

— — — — — — — — — — — — —

a. X   d. — average
b. X   e. X
c. X   f. —

84
Mrs. Allen received the "Three D" treatment, which consists of:
1.
2.
3.

— — — — — — — — — — — — — —

1. diet—low sodium
2. digitalization
3. diuretics

Review—Edema

Q. 1
Edema is the result of excess _____ whereas water intoxication

is the result of excess _____ .

— — — — — — — — — — — — —

sodium, water

Q. 2
Edema is the abnormal retention of fluid in the *_____

_____ . Without sodium excess, water (would/
would not) be retained.

— — — — — — — — — — — — — —

interstitial spaces or in serous cavities, would not

Q. 3

The five main physiological factors leading to edema are

1.

2.

3.

4.

5.

— — — — — — — — — — — — — —

1. increased hydrostatic pressure
2. decreased colloid osmotic pressure
3. increased capillary permeability
4. increased sodium retention
5. decreased lymphatic drainage

Q. 4

Will the venous (hydrostatic) pressure in an individual in congestive

heart failure (CHF) be increased or decreased? _____ .

Why? _____

— — — — — — — — — — — — — —

increased, The blood is backed in the venous system causing increased pressure.

Q. 5

In burns, there will be a(n) (increase/decrease) in the plasma colloid

osmotic pressure. Why? _____ .

— — — — — — — — — — — — —

decrease, There is a decrease in plasma protein.

Q. 6

With a decrease in plasma colloid osmotic pressure, water will flow from

the *_____ into the *_____ .

_ _ _ _ _ _ _ _ _ _ _ _ _ _ _

intravascular spaces, into the interstitial spaces

Q. 7

Generalized edema is called _____ .
   Can you recall the names of fluid found in the following body cavities?
If not, refer to Frame 71.
Peritoneal cavity: _____

Pleural cavity: _____

Pericardial cavity: _____

_ _ _ _ _ _ _ _ _ _ _ _ _ _ _

anascara, ascites, hydrothorax, hydropericardium

Q. 8

Individuals having poor cardiac output are prone to develop fluid

throughout the lung. This is known as *_____ .

_ _ _ _ _ _ _ _ _ _ _ _ _ _ _

pulmonary edema

Q. 9

Digitalis preparations will slow down the *_____

and make the heart beat more _____ . Will digitalis preparations increase or decrease cardiac and urinary output?

_ _ _ _ _ _ _ _ _ _ _ _ _ _ _

ventricular contractions, forcefully, increase

Q. 10

Diuretics are used to (increase/decrease) the secretion of urine.
  The thiazide diuretics will cause an excretion of sodium and water,

but also the valuable electrolyte _____.

– – – – – – – – – – – – – – –

increase, potassium

SHOCK

85

The state of circulatory collapse, known as shock, occurs when the hemostatic circulatory mechanism, which regulates circulation, fails to maintain adequate circulation. With shock, the cardiac output is insufficient
to provide vital organs and tissues with blood.

  Shock is a state of *_____.
  Shock occurs when the hemostatic circulatory mechanism fails to

*_____

_____.

– – – – – – – – – – – – – – –

circulatory collapse, maintain adequate circulation or provide adequate
blood to vital organs or tissues.

86

A common feature of shock, regardless of the cause, is a low blood volume in relation to the vascular capacity. There is a loss of blood, not necessarily from hemorrhaging, but from "pooling" in body areas so the blood does not circulate.

A low blood volume is known as _____.

A disproportion between the volume of blood and the capacity (size)

of the vascular chamber is the essential feature of _____

_____.

_ _ _ _ _ _ _ _ _ _ _ _ _ _ _

hypovolemia, shock or circulatory collapse

87

A common feature of shock is *_____.

With shock, is hypovolemia always due to hemorrhaging? _____.

Explain. _____.

_ _ _ _ _ _ _ _ _ _ _ _ _ _ _

low blood volume or loss of blood.
NO! Can be due to pooling of blood in body areas.

88

When the blood volume becomes too small for the vascular capacity,
venous blood return to the heart is reduced, and there is a drop in
cardiac output and systemic arterial blood pressure. This will lead to an
inadequate return of blood to the right side of the heart.

The cardiac output is dependent on:

( ) a. arterial return
( ) b. venous return

With a decrease in venous return, the amount of circulating blood

would be _____.

– – – – – – – – – – – – – – –

a. –
b. X
decreased

Table 25 outlines the physiological factors resulting from shock. It includes whether an increase or decrease of the physiological factor would
lead to shock. The rationale gives an explanation of the reasons, causes,
and results of shock according to the physiological factors.

Study this table carefully, noting whether it is an increase or a decrease
in the physiological factor which can lead to shock. Be able to give one
explanation for each from the rationale column. Refer to this table as
needed.

TABLE 25 The Physiological Factors Resulting from Shock

Physiological Factors		Rationale
Arterial blood pressure	↓ Decreased	A reduced venous return to the heart will decrease the cardiac output and arterial blood pressure (B.P.).
		A decrease in B.P. is sensed by the pressoreceptors in the carotid sinus and aortic arch which leads to an immediate reflex increase in systemic vasomotor activity. (This center is found in the medulla.) The result will be in cardiac acceleration and vasoconstriction in order to maintain homeostasis with respect to blood pressure. This may be sufficient for early or impending shock.
Vasoconstriction of the blood vessels	↑ Increased	Increased sympathetic nerve activity will cause vasoconstriction. Vasoconstriction will tend to maintain blood pressure and reduce the discrepancy between blood volume and vascular capacity (size). Vasoconstriction is greatest in the skin, kidneys, and skeletal muscles and not as significant in the cerebral vessels.
		The coronary arteries actually dilate with a decrease in blood volume. This will provide sufficient blood to the heart muscle (myocardium) for heart function.
Heart rate	↑ Increased	Heart rate is increased to overcome poor cardiac output and to increase circulation.
		A rapid, thready pulse will be one of the first identified signs of shock.

TABLE 25 (continued)

Physiological Factors	Rationale
Metabolic changes	↓ Decreased — The fall in plasma hydrostatic pressure reduces urinary filtration.
	Unopposed plasma colloidal osmotic pressure draws interstitial fluid into the vascular bed.
	Blood loss results in a loss of serum potassium, phosphate, and bicarbonate.
	Inadequate oxygenation of cells prevents their normal metabolism and leads to the formation of acid metabolities, thus lowering serum pH values. With a loss of K, HCO_3, and PO_4 and a fall in serum pH, metabolic acidosis results.
	A rise in blood sugar will first be seen due to the release of adrenalin; later, blood sugar will fall due to a decline in liver glycogen.
Kidney function	↓ Decreased — Low blood pressure causes inadequate circulation of blood which will result in renal ischemia which is a lack of O_2 to the kidney. Renal insufficiency may follow prolonged hypotension. The systolic blood pressure must be 60 mm Hg and above for maintaining kidney function.
	One of the body's compensatory mechanisms in shock is to shunt the blood around the kidney to maintain intravascular fluid. Deficient blood supply makes the tubule cells of the kidneys more susceptible to injury.
	Urine output of less than 25 ml per hour may be indicative of shock.

89

Place I for increase and D for decrease concerning the physiological factors as they occur with shock.

_____ a. Arterial blood pressure

_____ b. Kidney function

_____ c. Heart rate

_____ d. Metabolic changes

_____ e. Vasoconstriction

— — — — — — — — — — — — — — —

a. D   d. D
b. D   e. I
c. I

90

When there is a low blood pressure, the pressoreceptors in the carotid sinus and aortic arch will cause an increase in the systemic vasomotor activity which will lead to what two activities in order to maintain

homeostasis? _____ and _____.

Increased systemic vasomotor activity occurs in order to maintain

_____.

— — — — — — — — — — — — — — —

vasoconstriction and cardiac acceleration, homeostasis

91

Increased sympathetic activity will result in (vasoconstriction/vasodilatation).

Vasoconstriction is greatest in what three parts of the body? _____,

_____, and _____.

The coronary arteries will (dilate/constrict) with a decrease in blood volume.

- - - - - - - - - - - - - - -

vasoconstriction, skin, kidneys, and skeletal muscles, dilate

92

Heart rate in shock will be (increased/decreased) to overcome poor cardiac output and increase circulation.

The pulse rate would be _____ and _____.
A person with a pulse rate above 120 would have (bradycardia/tachycardia).

- - - - - - - - - - - - - -

increased, rapid and thready, tachycardia

93

The following metabolic changes would occur with shock:
Fluid would be drawn from the interstitial space into the vascular

space due to what kind of pressure? _____.

Inadequate oxygenation of cells will lead to the formation of acid metabolites, causing the pH to (rise/fall).

A fall in pH, loss of K, HCO_3, PO_4, will lead to (metabolic acidosis/metabolic alkalosis).

In shock, there will be a release of adrenalin which will cause the blood sugar to (rise/fall). Later, there will be a (rise/fall) due to a decline in liver glycogen.

- - - - - - - - - - - - - - -

colloidal osmotic pressure, fall, metabolic acidosis, rise, drop

94

In shock, the compensatory mechanisms will shunt the blood around

the kidney in order to maintain the volume of _____

This results in a lack of oxygen in the kidneys, known as _____

_____ , causing a decrease in kidney function.

The systolic blood pressure for kidney function must be at least _____

_____ .

An indication of shock related to kidney dysfunction would be a urine

output of less than _____ .

_ _ _ _ _ _ _ _ _ _ _ _ _ _ _

intravascular fluid, renal ischemia, 60 mm Hg, 25 ml per hour

Review—Physiological Factors of Shock

Place I for increased and D for decreased concerning the physiological
factors as they occur with shock. Give one explanation for each physiological factor. (Not all the answers have been given, so after you have
completed this review, check your answers against Table 25.)

_____Arterial blood pressure.

_____

_____

D. Reduced venous return will decrease cardiac output and arterial B.P.
or a decrease in B.P. will stimulate systemic vasomotor activity or others.

_____Vasoconstriction.

_____

_____

I. Will result from an increased sympathetic activity or vasoconstriction
will tend to raise the B.P. or others.

_____Heart rate.

_____

_____

I. Increased heart rate will tend to increase cardiac output and circulation.

_____Metabolic changes.

_____

_____

D. Inadequate oxygenation of cells will prevent their normal metabolism and lead to metabolic acidosis or refer to Table 25.

_____Kidney function.

_____

_____

D. Low blood pressure will result in renal ischemia and a decrease in urinary output or others.

Clinical Considerations

95

Normal blood pressure is the usual level of blood pressure in an individual, and will vary to some extent from person to person.

Therefore, a blood pressure of less than 100 mm Hg is significant of shock in most individuals.

A systolic pressure of 60 to 70 mm Hg is necessary to maintain the coronary circulation and renal function (urinary output). An individual

with a systolic pressure of 50–60 mm Hg is said to be in _____.

_ _ _ _ _ _ _ _ _ _ _ _ _ _ _ _

shock

96

A low pulse pressure, which is the difference between the systolic and diastolic pressures, is indicative of shock. To maintain coronary circulation and renal function, the systolic pressure should be at least

_____.

A pulse pressure of 20 mm Hg or less is indicative of _____.

_ _ _ _ _ _ _ _ _ _ _ _ _ _ _

60–70 mm Hg, shock

97

Blood supply to the organs most susceptible to acute anoxia (absence or lack of oxygen), i.e., the brain and the heart, is maintained as long as possible at the expense of the less vital organs and tissues.

The two organs most susceptible to anoxia are _____ and

_____.

The brain can survive 4 minutes in an anoxic state before cerebral damage occurs.

_ _ _ _ _ _ _ _ _ _ _ _ _ _ _

heart and brain

98

Which of the following would indicate shock?

( ) a. Arterial blood pressure of less than 100

( ) b. Pulse pressure of 50 mm Hg

( ) c. Pulse pressure of 20 mm Hg

The two organs most susceptible to acute anoxia are:

( ) a. Heart

( ) b. Brain

( ) c. Intestines

The organ which cannot survive anoxia longer than 4 minutes without

permanent damage is the _____.

– – – – – – – – – – – – – – –

a. X   a. X

b. –   b. X

c. X   c. –

brain

Table 26 describes four types of shock: hematogenic shock, also known as hemorrhagic shock; cardiogenic shock; septic shock, also known as endotoxic shock; and neurogenic shock. Clinical causes for each type of shock are also included.

Study this table carefully and be able to explain the rationale for each type of shock and be able to give at least one clinical contributing cause. Refer to this table as needed.

TABLE 26   Types of Shock

Types	Rationale	Clinical Causes
Hematogenic shock (hemorrhagic)	Hematogenic shock is due to loss of blood and plasma, creating a decrease in the circulating blood volume.	Hemorrhaging might result from surgery or a traumatic injury, or burns, or gastro-intestinal bleeding.
Cardiogenic shock	Cardiogenic shock is due to the failure of the pumping action of the myocardium creating a decrease in the circulating blood volume, thus causing an inadequacy in tissue perfusion. The venous pressure is elevated due to the accumulation of blood in the heart and large veins.	This can result from myocardial infarction, cardiac failure.
Septic shock (endotoxic)	Septic shock is characterized by capillary permeability permitting blood and plasma to pass into the surrounding tissues.	This can be due to a bacterial infection.
Neurogenic shock	Neurogenic shock is due to the loss of vascular tone. This leads to vasodilatation.	It can result from emotional factors, spinal anesthesia, or trauma from an extensive operative procedure.

99

The four types of shock are:
1.
2.
3.
4.

— — — — — — — — — — — — — —

1. hematogenic shock
2. cardiogenic shock
3. septic shock
4. neurogenic shock

100

Match the types of shock with the rationale.
a. Hematogenic shock
b. Cardiogenic shock
c. Septic shock
d. Neurogenic shock

_____ 1. Failure of the myocardium causing a decrease in the circulating blood volume.

_____ 2. Loss of vascular tone with vasodilatation.

_____ 3. Decrease in blood volume due to loss of blood and plasma.

_____ 4. Capillary permeability resulting from an infection.

_ _ _ _ _ _ _ _ _ _ _ _ _ _ _

1. b   3. a
2. d   4. c

101

Match the types of shock with the clinical causes.
a. Hematogenic shock
b. Cardiogenic shock
c. Septic shock
d. Neurogenic shock

_____ 1. Spinal anesthesia, emotional factors, or trauma from an extensive operative procedure.

_____ 2. Hemorrhaging from surgery or injury, burns, or gastrointestinal bleeding.

_____ 3. Severe bacterial infection.

_____ 4. Myocardial infarction and/or cardiac failure.

_ _ _ _ _ _ _ _ _ _ _ _ _ _ _

1. d   3. c
2. a   4. b

102

Complete the following table. Give at least one clinical cause for each type of shock. You may find it difficult to recall all the rationale, so refer to the original table if necessary

Types	Rationale	Clinical Causes
Hematogenic shock		
Cardiogenic shock		
Septic shock		
Neurogenic shock		

– – – – – – – – – – – – – – –

Hematogenic shock—Blood volume is decreased due to loss of blood plasma. Causes: Burns, hemorrhage, or gastrointestinal bleeding.

Cardiogenic shock—Failure of myocardium, decreasing the circulating blood volume. Causes: heart failure.

Septic shock—Capillary permeability decreasing vascular blood volume. Causes: bacterial infections.

Neurogenic shock—Loss of vascular tone. Vasodilatation. Causes: Emotional factors, spinal anesthesia, or extensive operative procedure.

Table 27 is a chart of the signs and symptoms of shock, placed in the order in which they frequently occur. You will be expected to memorize this chart and note that heart rate or pulse will change before the blood pressure decreases.

Refer to the Glossary for unfamiliar words and refer to this table as needed.

TABLE 27   Signs and Symptoms of Shock*

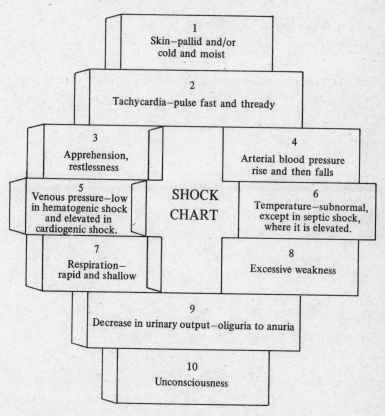

1
Skin—pallid and/or
cold and moist

2
Tachycardia—pulse fast and thready

3
Apprehension,
restlessness

4
Arterial blood pressure
rise and then falls

5
Venous pressure—low
in hematogenic shock
and elevated in
cardiogenic shock.

SHOCK
CHART

6
Temperature—subnormal,
except in septic shock,
where it is elevated.

7
Respiration—
rapid and shallow

8
Excessive weakness

9
Decrease in urinary output—oliguria to anuria

10
Unconsciousness

*These signs and symptoms are placed in the order in which they frequently occur.

103

In shock, tachycardia is frequently seen before the arterial blood pressure begins to fall.

The heart beats faster to (increase/decrease) the circulating blood volume. This is an early compensatory mechanism to overcome shock.

With shock, what happens to the arterial blood pressure? _____

_____.

_ _ _ _ _ _ _ _ _ _ _ _ _ _ _ _

increase, It will first rise and then fall.

104

The first three signs and symptoms of shock according to the order of frequency of occurrence are:

1.
2.
3.

_ _ _ _ _ _ _ _ _ _ _ _ _ _ _

1. skin—pallid and/or cold and moist
2. tachycardia
3. apprehension

105

Check the following signs and symptoms which are indicative of shock:
( ) a. Venous pressure low in cardiac shock
( ) b. Arterial blood pressure of 70 mm Hg
( ) c. Restlessness, apprehension
( ) d. Excess urine output
( ) e. Venous pressure low in hematogenic shock
( ) f. Temperature elevated in septic shock
( ) g. Bradycardia
( ) h. Tachycardia
( ) i. Respiration—rapid and shallow
( ) j. Skin is pallid and/or cold and moist

— — — — — — — — — — — — — — —

a. —	f. X
b. X	g. —
c. X	h. X
d. —	i. X
e. X	j. X

106

Below are some statements about signs and symptoms of shock. Check the true statements. Correct the false ones.
( ) a. Arterial blood pressure is low.
( ) b. Bradycardia
( ) c. Respiration—slow and deep
( ) d. Apprehension, restlessness
( ) e. Temperature low in all types of shock
( ) f. Increased urine output
( ) g. Venous pressure is high in cardiac shock and in hematogenic shock
( ) h. Skin is pallid and hot

— — — — — — — — — — — — — — —

a. Arterial blood pressure rises, then falls	e. Not septic shock
	f. Decreased output
b. Tachycardia	g. It is low in hematogenic shock
c. Fast and shallow	h. Pallid and/or cold and moist
d. X	

Clinical Management

107

The four clinical causes of hematogenic shock include:

1.
2.
3.
4.

— — — — — — — — — — — — — — —

1. surgery
2. traumatic injury
3. burns
4. gastrointestinal bleeding

Table 28 outlines the calculation of blood loss and fluid replacement for three states of hematogenic shock. With hematogenic shock, there is an actual blood loss from the body; therefore, body fluid needs to be replaced.

In planning parenteral replacement therapy for patients in shock, factors responsible for the conditions are first considered. Factors responsible are frequently referred to as causes or causative agents.

Study this table very carefully, noting the states of hematogenic shock, the systolic blood pressure ranges, the estimated blood volume loss according to an individual weighing 70 Kgs, and the fluid replacement needed. Refer to this table whenever it is necessary.

**TABLE 28 Calculation of Blood Loss and Fluid Replacement
for Three States of Hematogenic Shock**

State of Hematogenic Shock	Systolic Blood Pressure (mm Hg)	Estimated Blood Volume Loss in Percent	Replacement Needed for Blood Volume Loss in an Individual Weighing 70 Kg
Mild shock	95	15–20%	750–1100 cc (ml)
Moderate shock	80 to 95	20–30%	1100–1700 cc (ml)
Severe shock	Below 80	30–50%	2000–3000 cc (ml) and up

Note:
When blood is not available, large amounts of balanced salt solution, 100–150 cc/Kg of body weight, are given.

There should be 50 cc replacement for each percent blood volume loss in a 70 Kg individual or as determined by the physician.

108

If a patient is said to be in <u>mild</u> hematogenic shock, his blood volume

loss is _____ %, and his systolic blood pressure is probably

_____ mm Hg. The replacement needed for this blood volume

loss is _____ cc.

If a patient is said to be in <u>moderate</u> hematogenic shock, his blood

volume loss is _____ %, and his systolic blood pressure is probably

_____ mm Hg. The replacement needed for this blood volume loss

is _____ cc.

Or if a patient is said to be in <u>severe</u> hematogenic shock, his blood vol-

ume loss is _____ %, and his systolic blood pressure is probably

_____ mm Hg. The replacement needed for this blood volume loss

is _____ cc.

– – – – – – – – – – – – – – – – –

15–20%, 95 mm Hg, 750–1100 cc, 20–30%, 80-95 mm Hg, 1100–1700 cc,
30–50%, below 80 mm Hg, 2000–3000 cc

109

There should be _____ cc replacement for each percent blood volume loss in a man of average weight, 150 pounds or _____ Kg.

_ _ _ _ _ _ _ _ _ _ _ _ _ _ _

50 cc, 70 Kg

110

Place the word Mild, Moderate, or Severe in the space provided as it relates:

_____ Systolic blood pressure below 80 mm Hg

_____ Systolic blood pressure 88 mm Hg

_____ Systolic blood pressure around 95 mm Hg

_____ 50% blood volume loss

_____ 15–20% blood volume loss

_____ 20–30% blood volume loss

_____ 1100–1700 cc replacement needed

_____ 2000–3000 cc replacement needed

_____ 750–1100 cc replacement needed

_ _ _ _ _ _ _ _ _ _ _ _ _ _ _

Severe   Moderate
Moderate   Moderate
Mild   Severe
Severe   Mild
Mild

Table 29 outlines the clinical management for alleviating four types of clinical shock: hematogenic, cardiogenic, septic, and neurogenic. Years ago the first and foremost treatment of shock was to administer a vasopressor drug. The drug would constrict the dilated blood vessels, which occur with shock, and would raise the blood pressure. Vasopressors will act as only a temporary treatment or solution to shock, and shock will continue to increase if the cause is not alleviated or removed. Today vasopressors are used for severe shock and types of shock nonresponsive to treatment. Note that vasopressors are not effective in the treatment of hematogenic shock for constricting blood vessels will not aid in the circulation of blood when the cause is most obvious—a lack of blood, causing hypovolemia. Replacing blood volume loss should correct this type of shock. Remember, removal of the cause is first and foremost in alleviating various types of shock.

Study this table most carefully and be able to explain the treatments for each type of shock. Refer to this table as needed.

TABLE 29 Clinical Management for Alleviating Various Types of Clinical Shock

Hematogenic Shock	Cardiogenic Shock	Septic Shock	Neurogenic Shock
1. Parenteral fluids, such as: a. Dextran solution b. Lactated Ringer's c. Normal saline d. Whole blood	1. Parenteral therapy is limited when edema is present and the venous pressure is elevated. A close monitoring of central venous pressure (CVP) is indicated.	1. Blood culture first, and then parenteral fluids.	1. Parenteral therapy is limited unless condition warrants it.
2. No vasopressors	*2. Vasopressors if severe	*2. Vasopressors for nonresponsiveness	*2. Vasopressors if severe
3. Electrolyte replacement	3. Digitalization, sedation, lidocaine as needed abort ventricular arrhythmias	3. Antibiotics via I.V. fluids Steroids, e.g., hydrocortisone	3. Sedation for emotional shock
4. Hyperbaric oxygenation— O_2 under pressure carried in the plasma	4. O_2		

*Examples of vasopressors are (1) Levarterenal bitartrate—Levophed; (2) Metaraminol bitartrate—Aramine.

111

What is shock? _____.

What is a common feature of shock? _____

_____.

_ _ _ _ _ _ _ _ _ _ _ _ _ _ _

State of circulatory collapse. Low blood volume or hypovolemia. This could be due to loss of blood from the body or "pooling" of blood.

112

The clinical management for hematogenic shock may consist of:

( ) a. Dextran I.V. solution
( ) b. Whole blood
( ) c. Digitalization
( ) d. Vasopressors
( ) e. O$_2$ (Oxygen)
( ) f. Lactated Ringer's solution
( ) g. Normal saline
( ) h. Electrolyte replacement
( ) i. Hyperbaric oxygenation
( ) j. Lidocaine

_ _ _ _ _ _ _ _ _ _ _ _ _ _ _

a. X   f. X
b. X   g. X
c. −   h. X
d. −   i. X
e. −   j. −

113

Clinical management for <u>cardiogenic shock</u> may consist of:

( ) a. Limited parenteral therapy
( ) b. No vasopressors
( ) c. Antibiotics
( ) d. O$_2$
( ) e. Digitalization
( ) f. Sedation

- - - - - - - - - - - - - - -

a. X d. X
b. − e. X
c. − f. X

114

Clinical management for septic shock may consist of:

( ) a. Blood culture
( ) b. Antibiotics in intravenous fluids
( ) c. Hydrocortisone
( ) d. Vasopressors
( ) e. Digitalization
( ) f. Massive parenteral therapy with whole blood

- - - - - - - - - - - - - - -

a. X d. X
b. X e. −
c. X f. −

115

Clinical management for neurogenic shock may consist of:

( ) a. Blood culture
( ) b. Vasopressors
( ) c. Limited parenteral therapy
( ) d. Massive parenteral therapy
( ) e. Sedation

– – – – – – – – – – – – – –

a. –   d. –
b. X   e. X
c. X

116

Explain the action of vasopressors. _____

_____ .

Give the trade names of two vasopressors _____

and _____ .

– – – – – – – – – – – – – –

Vasopressors constrict blood vessels in hopes of improving circulation.
Levophed and Aramine

117

Identify which of the treatments listed below might be used in various
types of shock by placing:

    H for hematogenic shock
    C for cardiogenic shock
    S for septic shock
    N for neurogenic shock

Some treatments may be used for more than one type of shock.

_____ Parenteral therapy—Dextran, lactated Ringer's, or normal
                saline

_____ Digitalization

_____ Electrolyte replacement

___, ___ Sedation

___, ___, ___Vasopressors

_____ Oxygen

___, ___ Limited parenteral therapy

_____ Hyperbaric oxygenation

_____ Antibiotics in intravenous fluids

_____ Hydrocortisone

- - - - - - - - - - - - - - -

H	C
C	C, N
H	H
C, N	S
C, S, N	S

Review—Shock

Q. 1

Shock occurs when the hemostatic circulatory mechanism fails to

maintain adequate _____.

— — — — — — — — — — — — —

circulation or blood volume

Q. 2

The common feature of different types of shock is the low blood

volume in relation to the _____.

— — — — — — — — — — — — —

vascular capacity or size

Q. 3

A decrease in blood volume is known as _____.

— — — — — — — — — — — — —

hypovolemia

Q. 4

A reduced venous blood return to the heart will result in a decrease

in *_____ and a decrease in *_____

_____.

— — — — — — — — — — — — —

cardiac output, blood pressure

Q. 5

Increase in the systemic vasomotor activity will lead to cardiac

_____ and _____.

— — — — — — — — — — — — —

acceleration and vasoconstriction

Q. 6
In shock, the heart rate will _____ .

- - - - - - - - - - - - - -

increase

Q. 7
Inadequate oxygenation of cells will prevent their normal metabolism and lead to metabolic (acidosis/alkalosis).

- - - - - - - - - - - - - -

acidosis

Q. 8
Low blood pressure will result in renal ischemia. Will this cause urine

output to increase or decrease? _____ .

- - - - - - - - - - - - - -

decrease

Q. 9
How is pulse pressure determined? _____ .

What might a low pulse pressure indicate? _____ .

- - - - - - - - - - - - - -

the difference between the systolic and diastolic pressure. shock

---

Q. 10
The organ of the body which cannot survive from acute anoxia after

4 minutes is the _____ .

- - - - - - - - - - - - - -

brain

Q. 11
Hematogenic shock is due to loss of _____ , whereas
neurogenic shock is due to a loss of * _____ .

_ _ _ _ _ _ _ _ _ _ _ _ _ _ _ _

blood, vascular tone

Q. 12
In shock, tachycardia is frequently seen before the _____
_____ begins to fall.

_ _ _ _ _ _ _ _ _ _ _ _ _ _ _

arterial blood pressure

Q. 13
The venous pressure in hematogenic shock is _____ and in
cardiogenic shock is _____ .

_ _ _ _ _ _ _ _ _ _ _ _ _ _ _

decreased, increased

Q. 14
Clinical management for hematogenic shock may consist of:
1.
2.
3.
4.

_ _ _ _ _ _ _ _ _ _ _ _ _ _ _

1. Dextran 6% in saline or lactated Ringer's or normal saline or whole
   blood
2. No vasopressors
3. Electrolyte replacement
4. Hyperbaric oxygenation

Q. 15

Clinical management for cardiogenic shock may consist of:
1.
2.
3.
4.

– – – – – – – – – – – – – –

1. Limited parenteral fluids with edema
2. Digitalization
3. Vasopressors
4. Oxygen
5. Sedation

SUMMARY QUESTIONS

The questions in this summary will test your comprehensive ability and knowledge of the material found in Chapter 5. If the answer is unknown, refer to the section in the chapter for further understanding and clarification.

1. Define dehydration and hypertonic dehydration.
2. What are the three degrees of dehydration and differentiate between the three in reference to body weight loss, symptoms, and body water deficit.
3. Explain why hemoconcentration occurs with dehydration.
4. How does the balance of intracellular and extracellular fluid loss change between early dehydration and chronic dehydration?
5. The clinical management for dehydration is to replace body water loss. The total fluid deficit is estimated according to the percent of

   _____. One-third of body

   water deficit is from the _____ fluid and two-thirds

   of body water deficit is from the _____ fluid.
6. Give the suggested solution replacement for correcting dehydration.
7. Define water intoxication and explain how it can occur.
8. Differentiate between water intoxication and edema, emphasizing what happens to the cells in each case.

9. What are the early symptoms of water intoxication?
10. Give the overall objective for the clinical management of water intoxication and name two ways in which this objective can be accomplished.
11. What are the three methods for promoting water excretion?
12. Place the word increased or decreased concerning the physiological factors as they occur in edema:

_____ Hydrostatic pressure in the capillaries

_____ Plasma colloid osmotic pressure

_____ Capillary permeability

_____ Sodium retention

_____ Lymphatic drainage

Explain the rationale for the above physiological factors as they relate to edema.
13. Name several clinical situations in which edema is found.
14. What are the effects of excessive parenteral fluids on an individual in severe congestive heart failure?
15. What is hypervolemia?
16. Explain the "Three D's" in the clinical management of edema. What imbalance should the nurse be observing when the patient is receiving a daily digitalis preparation and a daily diuretic (thiazide)?
17. Shock has been defined as the state of _____

_____. Explain the five physiological factors resulting from shock.
18. What is hypovolemia? The essential feature of shock is the disproportion between the _____ and the _____.
19. What is the systolic pressure necessary to maintain coronary circulation and renal function? What organ cannot survive anoxia longer than 4 minutes without damage?
20. Define the types of shock and give examples: hematogenic shock, cardiogenic shock, septic shock, neurogenic shock.
21. What are the symptoms of shock? Which three are the most commonly seen?
22. Explain the clinical management for alleviating the four types of clinical shock.

References

Beland, Irene L., *Clinical Nursing: Pathophysiological and Psychosocial Approaches*, New York: Macmillan Co., 1965, pp. 605–608.

Brunner, Lillian Sholtis, *et al.*, *Medical-Surgical Nursing*, Philadelphia, Pa.: J. B. Lippincott Co., 1964, pp. 162–167.

Burgess, Richard, "Fluids and Electrolytes," *American Journal of Nursing*, LXV (October, 1965), pp. 94–95.

Grollman, Arthur, "Diuretics," *American Journal of Nursing*, LXV (January, 1965), pp. 84–89.

Guyton, Arthur C., *Function of the Human Body*, 2nd ed., Philadelphia, Pa.: W. B. Saunders Co., 1964, pp. 171–174.

Guyton, Arthur C., *Textbook of Medical Physiology*, 3rd ed., Philadelphia, Pa.: W. B. Saunders Co., 1966, pp. 451–457, 540–541.

Harrison, T. R., ed., *Principles of Internal Medicine*, 4th ed., New York: McGraw-Hill Book Co., 1962, pp. 1380–1391.

Jacob, Stanley W. and Clarice A. Francone, *Structure and Function in Man*, Philadelphia, Pa.: W. B. Saunders Co., 1965, pp. 461–462.

Jenkinson, Viven M., "Congestive Heart Failure," *Basic Medical-Surgical Nursing*, Dubuque, Iowa: William C. Brown Co., 1966, pp. 166–171.

Krug, Elsie E., *Pharmacology in Nursing*, 9th ed., St. Louis, Mo.: The C. V. Mosby Co., 1963, pp. 517–528.

Metheney, Norma M. and William D. Snively, *Nurses' Handbook of Fluid Balance*, Philadelphia, Pa.: J. B. Lippincott Co., 1967, pp. 199–204.

Morris, Dona G., "The Patient in Cardiogenic Shock," *Cardio-Vascular Nursing*, 5 (July–August, 1969), pp. 15–17.

Musser, Ruth D. and Betty Lou Shubkagel, *Pharmacology and Therapeutics*, 3rd ed., New York: The Macmillan Co., 1965, pp. 629–641.

Shafer, Kathleen N., *et al.*, *Medical-Surgical Nursing*, 3rd ed., St. Louis, Mo.: The C. V. Mosby Co., 1964, pp. 239–250.

"Shock," *Hospital Focus* (October 1, 1962), 16 pp. (Knoll Pharma. Co.).

Simeone, F. A., "Shock: It's Nature and Treatment," *American Journal of Nursing*, LXVL (June, 1966), pp. 1286–1294.

Smith, Dorothy W. and Claudia D. Gips, *Care of the Adult Patient*, Philadelphia, Pa.: J. B. Lippincott Co., 1963, pp. 707–723.

Statland, Harry, *Fluid and Electrolytes in Practice*, 3rd ed., Philadelphia, Pa.: J. B. Lippincott Co., 1963, pp. 76–83, 176–189.

Strickland, William M., "Replacement Therapy in Traumatic Shock," *Seminar Report*, VI (Spring, 1961), pp. 2–7 (Merck Sharp and Dohme).

Chapter 6

Clinical Situations—Gastrointestinal Surgery, Cirrhosis, Renal Failure, Burns, and Diabetic Acidosis

BEHAVIORAL OBJECTIVES

Upon completion of this chapter, the student will be able to:

Explain the physiological factors leading to fluid and electrolyte changes in selected clinical situations—gastrointestinal surgery, cirrhosis, renal failure, burns, and diabetic acidosis.

Assess fluid and electrolyte changes in given clinical examples and in actual clinical situations in a hospital or a community setting.

Apply clinical management in given clinical situations.

Plan nursing interventions to meet patient's needs in a given clinical situation. (You may need some assistance in planning nursing interventions.)

Name the adverse reactions to be observed in parenteral and drug replacement therapy for the patient in diabetic acidosis.

INTRODUCTION

In a clinical setting, the nurse will frequently provide care for individuals having fluid and electrolyte imbalance resulting from selected clinical situations. We will discuss five such situations: gastrointestinal surgery, cirrhosis, renal failure (corrected by peritoneal dialysis), burns, and diabetic acidosis. To assess the patients' needs and to provide the care needed for individuals with disease entities, the nurse must have the knowledge and understanding of fluid and electrolyte balance and imbalance. Through her knowledge and understanding, the nurse can then assess the changes occurring with her patients and plan the nursing interventions to meet these changes.

In this chapter, the participant will become acquainted with five individuals having various fluid and electrolyte imbalance resulting from

disease entities. The participant in this program will first gain an understanding of the physiological factors involved in each clinical situation.
Clinical considerations, examples of representative patients, and clinical
management will follow. As a result of the clinical situations presented,
the nurse should be more cognizant of the fluid and electrolyte changes
occurring with her patients, and could then apply the knowledge and
understanding gained to other clinical situations.

GASTROINTESTINAL SURGERY

Physiological Factors

1

Individuals undergoing minor surgery frequently have little or no fluid
and electrolyte changes.

In major surgery, there is a tendency for sodium and water to be retained and for potassium to be lost. Before administering potassium, it
is necessary to make certain that:

( ) a. the individual can tolerate food
( ) b. renal function is adequate

— — — — — — — — — — — — — — —

a. —
b. X

_____

2

Following major surgery, there is a tendency for sodium _____

_____, water _____and

potassium _____.

— — — — — — — — — — — — — — —

retention, retention, loss

3

Many individuals undergoing gastrointestinal surgery will have had a previous fluid and electrolyte imbalance along with concurrent losses. Replacement of losses is necessary, before, during, and following surgery.

Frequently, these individuals are dehydrated and will need sufficient water to reestablish renal function. The following type of solution would be indicated in reestablishing renal function:

( ) a. Hydrating solutions
( ) b. Plasma expanders
( ) c. Replacement solutions with potassium replacement

– – – – – – – – – – – – – – – –

a. X, b. –, c. –

4

Many individuals undergoing gastrointestinal surgery will require fluid and electrolyte replacement therapy:

( ) a. before surgery
( ) b. during surgery
( ) c. after surgery

– – – – – – – – – – – – – – – –

a. X, b. X, c. X

5

Gastric or intestinal intubation (tube passed into the stomach or intestine) for suctioning purposes will frequently be inserted before surgery. This will be used to alleviate vomiting due to an obstruction in the gastrointestinal tract or to decompress the stomach and/or bowel before and after an operation.

For gastric intubation, a Levine tube is inserted via nose into the stomach. For intestinal intubation, a Miller-Abbott tube or Cantor tube is inserted via nose and stomach into the intestine. The intestinal tubes are longer than the gastric tube, and they contain, on the tip of their tubes, a small balloon filled with air or mercury which aids in stimulating peristalsis (bowel tone). A gastric or an intestinal tube is frequently used following abdominal surgery in order to remove secretions until peristalsis returns and to relieve abdominal distention.

Gastric or intestinal intubation prior to abdominal surgery is used to
*
_____

_____ .

Gastric or intestinal intubation following abdominal surgery is em-

ployed to *_____

and to _____ .

– – – – – – – – – – – – – – –

alleviate vomiting and decompress the bowel or stomach.
remove secretions until peristalsis returns and to relieve abdominal
distension

Clinical Example

6

Mr. Drum, a 29-year-old man, was admitted to the hospital complaining of severe, persistent hiccups and abdominal pain. The patient noticed having a mass in the lower left quadrant of the abdomen for approximately 5-6 days before admission. According to Mr. Drum, he previously had several episodes of this left groin mass which he was able to reduce manually.

Mr. Drum stated he had not had a bowel movement for the past 3 days, nor was he able to "keep anything down" over the past 3 days. His skin was warm and dry, and lacked elasticity. He was very weak.

Diagnoses

1. Incarcerated left inguinal hernia with probable small bowel obstruction.
2. Extreme dehydration.
3. Right inguinal hernia.

The following signs and symptoms might indicate that Mr. Drum was dehydrated:

( ) a. Vomiting—unable to retain food for 3 days
( ) b. Skin warm, dry and lacking elasticity
( ) c. Not having a bowel movement for 3 days
( ) d. Weakness
( ) e. Hernia could be manually reduced
( ) f. Severe abdominal pain in left lower quadrant

— — — — — — — — — — — — — —

a. X
b. X
c. —
d. X
e. —
f. —

TABLE 30   Laboratory Studies of Mr. Drum

Laboratory Tests*	On Admission	First Day	Second Day	Third Day	Fourth Day
Hematology Hemoglobin (12.9–17.0 grams)	21.2	18.4	13.1	13.2	
Hematocrit (40–46%)	58	54	38	39	
W. B. C. (white blood count) (5,000–10,000 cu mm)	10,700				
Biochemistry B. U. N. (blood urea nitrogen) (10–15 mg/100 ml)	85	68	68	19	19
Plasma CO_2 (50–70 vol %) (22–32 mEq/L)	52 / 24	61 / 28	–	39 / 18	50 / 22
Plasma chloride (98–107 mEq/L)	73	78	73	91	97
Plasma sodium (135–146 mEq/L)	122	128	122	132	145
Plasma potassium (3.5–5.3 mEq/L)	5.2	4.0	4.0	4.2	4.1

*Laboratory norms from the Wilmington Medical Center, Delaware Division.

Table 30 gives the laboratory studies of Mr. Drum and shows how his results deviate from the norms at the time of his illness. It is important for you to memorize the "normal" laboratory ranges as given in Table 30 in the left-hand column. You will be using these ranges throughout the chapter. However, you may refer to Table 30 as needed.

7

Give the "normal" ranges for hemoglobin, _____,

for hematocrit, _____, and for white blood count, _____.

_ _ _ _ _ _ _ _ _ _ _ _ _ _ _

12.9–17.0 grams
40–46%, 5,000–10,000 cu mm

8

The B.U.N. is the abbreviation for blood urea nitrogen. Do you know

how urea is formed? _____.

And how it is excreted? _____.

What is the "normal" B.U.N. range? _____.

_ _ _ _ _ _ _ _ _ _ _ _ _ _ _

As a by-product of protein metabolism.
Through the kidney.
10–15 mg/100 ml

9

The "normal" plasma CO_2 range is _____ vol. % and

_____ mEq/L.

Would a patient with a plasma CO_2 of 18 mEq/L be in metabolic
(acidosis/alkalosis)? Refer to Chapter 3 for further clarification.

_ _ _ _ _ _ _ _ _ _ _ _ _ _ _

50–70%
22–32 mEq/L
acidosis

10

The "normal" range for plasma chloride is 98–107 mEq/L. Do you recall the "normal" range for plasma, also referred to as serum, potassium, and sodium? Potassium _____ and sodium _____. Refer to Chapter II for further clarification.

– – – – – – – – – – – – – –

K: 3.5–5.3 mEq/L, Na: 135–146 mEq/L

11

Complete the chart giving the "normal" ranges for the following hematology and biochemistry laboratory tests:

Hematology
Hemoglobin:

_____ grams
Hematocrit:

_____ %
W.B.C.

_____ cu. mm.

Biochemistry
B.U.N.:

_____ mg/100 ml
Plasma CO_2 :

_____ vol %

_____ mEq/L
Plasma Cl: _____ mEq/L

Plasma Na: _____ mEq/L

Plasma K: _____ mEq/L

– – – – – – – – – – – – – –

Hematology
Hemoglobin: 12.9–17.0 grams
Hematocrit: 40–46%
W.B.C.: 5,000–10,000 cu mm

Biochemistry
B.U.N.: 10–15 mg/100 ml
Plasma CO_2: 50–70 vol %
       22–32 mEq/L
Plasma Cl: 98–107 mEq/L
   Na: 135–146 mEq/L
   K: 3.5–5.3 mEq/L

12

Which of the following admission laboratory results of Mr. Drum would indicate fluid and electrolyte imbalance:

( )   a. Hemoglobin 21.2 grams
( )   b. Hematocrit 58%
( )   c. Plasma CO_2 52%
( )   d. Plasma CO_2 24 mEq/L
( )   e. Plasma chloride 73 mEq/L
( )   f. Plasma sodium 122 mEq/L
( )   g. Plasma potassium 5.2 mEq/L

– – – – – – – – – – – – – – –

a. X, b. X, c. –, d. –, e. X, f. X, g. –

13

Mr. Drum's elevated hemoglobin and hematocrit on admission and the

first day postoperatively would indicate _____.

– – – – – – – – – – – – – –

dehydration–Did you answer hemoconcentration? OK

14

A high B.U.N. is indicative of renal impairment and/or dehydration. Mr. Drum's elevated B.U.N. would indicate:

( )   a. an increased urine output.
( )   b. a retention of urea, the by-product of protein metabolism, in the circulating blood.
( )   c. an abnormal excretion of urea, the by-product of protein metabolism.

– – – – – – – – – – – – – – –

a. –, b. X, c. –

15

The third day postoperatively, Mr. Drum's plasma CO_2 decreased. This would indicate:

( ) a. an increased bicarbonate ion in the plasma.
( ) b. a decreased bicarbonate ion in the plasma.
( ) c. metabolic acidosis.
( ) d. metabolic alkalosis.

_ _ _ _ _ _ _ _ _ _ _ _ _ _ _ _

a. —, b. X, c. X, d. —

16

Below are some laboratory results for Mr. Drum. Label which imbalance they might indicate, using:

D for dehydration
K for kidney dysfunction
E for electrolyte imbalance
N for metabolic acidosis
O for normal range or for those that do not pertain to the above four

Some results may be associated with more than one imbalance.

———— a. Hematocrit 38%

———— b. Hemoglobin 21.2 grams

———— c. B.U.N. 68 mEq/L

———— d. B.U.N. 19 mEq/L

———— e. Plasma potassium 4.0 mEq/L

———— f. Plasma sodium 122 mEq/L

———— g. Plasma CO_2 18 mEq/L

———— h. Plasma CO_2 28 mEq/L

———— i. Plasma chloride 73 mEq/L

_ _ _ _ _ _ _ _ _ _ _ _ _ _ _ _

a. O, b. D, c. K, D, d. O, e. O, f. E, g. M, h. O, i. E

Clinical Management—Preoperative

17

The preoperative management for Mr. Drum would include:

1. Hydrate rapidly utilizing 4 to 5 liters over the next 6–8 hours.
2. Insert a Levine tube and place suction gauge to low.
3. Prepare for O.R. for a left inguinal herniorrhaphy as soon as he is hydrated.

Solution for Hydration: 4500 cc 5% D /–1/2 NS (dextrose in 1/2 normal saline or 0.45%)

Due to vomiting, Mr. Drum had:

( ) a. severe dehydration.
( ) b. water intoxication.
( ) c. a loss of sodium and chloride
( ) d. a low serum bicarbonate level.
( ) e. a low serum potassium level.

He was hydrated (before/after) the herniorraphy.

A Levine tube was inserted to:

(.) a. relieve distention.
( ) b. remove secretions from the stomach.
( ) c. lessen vomiting.
( ) d. provide nutrition.

— — — — — — — — — — — — — —

a. X   before   a. X
b. –   b. X
c. X   c. X
d. X   d. –
e. –

Clinical Management—Postoperative

18

The postoperative management for Mr. Drum would include:

1. Connect Levine tube to low suction and check drainage hourly.
2. Monitor parenteral therapy:
   1000 cc 5% D / 1/2 NS
   1000 cc 5% D/NS with one ampul of sodium bicarbonate
   1000 cc 5% lactated Ringer's
3. Check urine hourly and test for specific gravity and pH.
4. Administer penicillin for febrile condition.
5. Others not related to fluid and electrolyte imbalance.

Mr. Drum received gastric intubation following surgery to *_____

_____ and to _____

_____.

_ _ _ _ _ _ _ _ _ _ _ _ _ _ _ _

relieve abdominal distention and to remove gastric secretions

_____

19

Gastrointestinal secretions contain solid particles that may accumulate
and obstruct the tube. Irrigation of the tube will assure patency and
proper drainage.

Frequent irrigations, using large amounts of water, should be avoided
to prevent the loss of fluid and electrolytes.

Irrigation of Mr. Drum's tube will assure *_____

_____.

What might result from frequent irrigations with a large quantity of

water? _____

_____.

_ _ _ _ _ _ _ _ _ _ _ _ _ _ _ _

patency to ensure proper drainage
It will wash the electrolytes out of the GI tract.

_____

20

The tube should be irrigated at specific intervals with small amounts of saline to keep it patent.

Change of position helps in alleviating obstruction and aids in maintaining patency of the tube.

The use of small amounts of air to check the patency of the tube may be ordered instead of solution in order to prevent the loss of fluid and electrolytes. One would listen with the stethoscope for a "woosh" sound.

The three methods that could be employed to check the patency of Mr. Drum's gastric tube are to:

1. _____

   _____ .

2. _____ .

3. _____

   _____

— — — — — — — — — — — — — — —

Irrigate at specific intervals with small amounts of saline.
Change of position.
Introduce small amount of air and listen with a stethoscope for a "woosh" sound.

21

Mr. Drum was allowed sips of water to alleviate the dryness in his mouth and lessen irritation in his throat.

Special attention should be taken to limit the amount of water by mouth for water can dilute the electrolytes found in the stomach and the suction would remove them.

What might happen to the electrolytes if Mr. Drum drank a great deal of water during intubation? _____

_____ .

— — — — — — — — — — — — — — —

The electrolytes in the stomach would be diluted and suction would remove them.

22

After Mr. Drum's Levine tube is removed, the nurse should observe him for:
1. feeling of fullness
2. vomiting
3. abdominal distention

This would indicate that Mr. Drum's gastrointestinal tract (is/is not) functioning.

The signs and symptoms which would indicate that Mr. Drum's peristalsis had not returned would be:

1. _____

2. _____

3. _____

– – – – – – – – – – – – – – –

is not   2. vomiting
1. feeling of fullness   3. abdominal distention

23

Frequently, the tube is clamped for a period of time and then unclamped. The amount of fluid is measured. A large amount of fluid returned by unclamped tube would indicate that peristalsis has not returned.

Feeling of fullness, vomiting, and abdominal distention are signs and symptoms that _____ has not returned.

Using the clamping and unclamping method, how would you know if peristalsis was present? _____

_____.

– – – – – – – – – – – – – –

peristalsis
A small amount of fluid return or none.

24

Because suction will remove fluids and electrolytes, oral fluid intake is restricted and parenteral therapy is used.

Mr. Drum received intravenous fluids containing dextrose, saline, and lactated Ringer's for:

( ) a. replacing sodium and chloride loss.
( ) b. maintaining nutritional needs.
( ) c. maintaining electrolyte balance.
( ) d. replacing and maintaining fluids volume.

‒ ‒ ‒ ‒ ‒ ‒ ‒ ‒ ‒ ‒ ‒ ‒ ‒ ‒ ‒

a. X, b. X, c. X, d. X

25

Mr. Drum received an ampul of sodium bicarbonate in one of his intravenous fluids. This would:

( ) a. increase the plasma CO_2 or bicarbonate.
( ) b. decrease the plasma CO_2 or bicarbonate.
( ) c. reduce his metabolic acidotic state.
( ) d. reduce his metabolic alkalotic state.

‒ ‒ ‒ ‒ ‒ ‒ ‒ ‒ ‒ ‒ ‒ ‒ ‒ ‒ ‒

a. X, b. −, c. X, d. −

Review—Gastrointestinal Surgery

Q. 1

In major surgery, there is a tendency for sodium and water to be

_____ and potassium to be _____.

‒ ‒ ‒ ‒ ‒ ‒ ‒ ‒ ‒ ‒ ‒ ‒ ‒ ‒

retained
excreted

Q. 2

Gastrointestinal intubation with low suction is frequently used pre-operatively to *_____

and to _____

_____ .

– – – – – – – – – – – – – – –

alleviate vomiting,
decompress the stomach or bowel

Q. 3

Mr. Drum's elevated hemoglobin and hematocrit on admission and the
first day postoperatively was indicative of _____ .

– – – – – – – – – – – – – – –

dehydration

Q. 4

What might Mr. Drum's elevated B.U.N. indicate? _____

_____ .

An elevated B.U.N. is also indicative of _____

and _____ .

– – – – – – – – – – – – – – – .

An abnormal retention of urea in the circulating blood.
renal impairment and dehydration

Q. 5

Mr. Drum's decreased plasma CO_2 would indicate _____

_____ .

– – – – – – – – – – – – – –

metabolic acidosis

Q. 6

Due to vomiting, Mr. Drum had a low serum level of what three ions?

_____ , _____ , and _____ .

_ _ _ _ _ _ _ _ _ _ _ _ _ _ _

sodium, chloride, and bicarbonate;
potassium could also be lost.

Q. 7

Frequent irrigations of the Levine tube with a large quantity of water

might have what effect? _____

_____ .

_ _ _ _ _ _ _ _ _ _ _ _ _ _ _

depletion of electrolytes in the GI tract.

Q. 8

The three methods used to check the patency of the gastric tube are to:

1. _____

   _____

2. _____

3. _____

_ _ _ _ _ _ _ _ _ _ _ _ _ _

irrigate at specific intervals with small amounts of saline.
change body position.
introduce small amount of air in tube.

Q. 9

Mr. Drum received intravenous fluids containing dextrose, saline, and lactated Ringer's for:

1. replacing _____

2. maintaining _____

3. maintaining _____

4. replacing _____

– – – – – – – – – – – – – – –

1. replacing Na and Cl loss.
2. maintaining nutritional needs.
3. maintaining electrolyte balance.
4. replacing and maintaining fluids volume.

CIRRHOSIS OF THE LIVER

Physiological Factors

26

Liver (hepatic) disease is frequently associated with sodium and water retention caused by increased portal pressure and increased aldosterone secretion.

Water is retained in excess to sodium. With hepatic disease, there is a

sodium and water retention caused by *_____

_____ and _____.

– – – – – – – – – – – – – – –

increased portal pressure and increased aldosterone

27.

Water is retained in _____ to sodium.

– – – – – – – – – – – – – – –

excess

28

Aldosterone (an adrenal cortical hormone) has a sodium- and water-retaining effect and a potassium-excreting effect.

. With an increase of aldosterone secretion, the serum potassium would

be _____.

_ _ _ _ _ _ _ _ _ _ _ _ _ _ _

decreased

29

In spite of the presence of an excess of total body sodium, hyponatre-

mia results because water is retained in *_____

_____ .

_ _ _ _ _ _ _ _ _ _ _ _ _ _ _

excess to sodium

30

Other contributing causes of hyponatremia are:

a. Sodium moves into the intracellular space replacing potassium which leaves the cells because of dehydration, malnutrition, and/or diuresis.
b. Prolonged use of potent diuretics.
c. Low (or restricted) sodium diet.

Hyponatremia frequently occurs with liver dysfunction because: water is retained in excess to sodium; sodium shifts into the intracellular

spaces; prolonged use of *_____ and/or

*_____.

_ _ _ _ _ _ _ _ _ _ _ _ _ _

potent diuretics and low sodium diet

31

The four contributing causes of hyponatremia are:

1. _____

2. _____

3. _____

4. _____

— — — — — — — — — — — — — —

1. Water is retained in excess to sodium.
2. Sodium moves into the intracellular spaces.
3. Prolonged use of potent diuretics.
4. Low sodium diet.

32

In addition to potassium loss resulting from increased aldosterone secre-

tion, dehydration, and _____,
diuresis leads to further potassium loss.

— — — — — — — — — — — — — — — —

malnutrition

33

In Chapter 5, in the section on Edema, ascites was defined. Can you re-

call its meaning? _____

_____ .

— — — — — — — — — — — — — —

An accumulation of fluid in the peritoneal cavity (abdomen).

34

With cirrhosis, ascites may develop suddenly or insidiously.

Ascites is *_____

_____.

— — — — — — — — — — — — — — —

an accumulation of fluid in the peritoneal cavity.

One of the major complications of cirrhosis of the liver is the development of ascites. It occurs frequently with cellular liver damage and portal hypertension. The portal circulation, which is the liver's circulatory system, becomes affected when there is cellular liver damage. The blood cannot circulate through the liver sufficiently, thus the portal pressure is greatly increased.

Since ascites is a "big" factor in fluid and electrolyte imbalance and since it frequently accompanies severe cirrhosis of the liver, emphasis will be placed on ascites.

Study Table 31 carefully, noting where there is an increase or decrease in the physiological factors resulting in ascites, and the rationale for these physiological factors. Refer to this table as needed.

TABLE 31 Pathogenesis of Ascites

Physiological Factors		Rationale
Portal obstruction and hypertension	Increased ↑	Portal obstruction resulting in portal vein hypertension will not of itself cause ascites. When ascites does accompany portal hypertension, its presence can be explained by the associated liver damage. Surgical relief of portal hypertension will relieve ascites without its fluid accumulating elsewhere in the body. Thus, portal hypertension influences fluid accumulation in the abdomen, but the fundamental cause of ascites lies in damage to the cellular structure of the liver.
Capillary permeability	Increased ↑	The capillary permeability is increased due to a permeability defect of capillary endothelium, which contributes to the transudation of fluid from the portal system into the abdomen.
Plasma osmotic pressure	Decreased ↓	With increased liver congestion and failure of the liver to synthesize albumin, protein-rich fluid will leave the capillaries and pass into the abdominal cavity, thus lowering the plasma osmotic pressure. (Hypoproteinemia and hypoalbuminema results.)
Retention of sodium and water	Increased ↑	When plasma volume is reduced, an increased hormonal response (aldosterone) occurs which will decrease urinary output and will cause retention of sodium and water.

35

Place I for increased and D for decreased concerning the physiological factors associated with ascites.

_____ 1. Portal obstruction and hypertension

_____ 2. Capillary permeability

_____ 3. Plasma osmotic pressure

_____ 4. Retention of sodium and water

_ _ _ _ _ _ _ _ _ _ _ _ _ _ _ _

1. I, 2. I, 2. D, 4. I

36

Portal hypertension influences fluid accumulation in the abdomen, but the fundamental cause of ascites lies in *_____

_____.

_ _ _ _ _ _ _ _ _ _ _,_ _ _ _ _

damage to the cellular structure of the liver.

37

With ascites, there is a permeability defect of capillary endothelium which contributes to *_____

_____.

Will permeability increase or decrease? _____.

_ _ _ _ _ _ _ _ _ _ _ _ _ _

the transudation of fluid.
increase

38

Because of increased liver congestion and failure of the liver to synthesize albumin, rich protein fluid will leave the _____ and pass into the * _____.

Will the plasma osmotic pressure be increased or decreased? _____

_____.

The result will be (hypoproteinemia/hyperproteinemia).

— — — — — — — — — — — — — —

capillaries, abdominal cavity
decreased
hypoproteinemia

39

The hormonal response (aldosterone) will cause:

a. _____

b. _____

— — — — — — — — — — — — — —

a. decrease in urinary output; b. retention of sodium and water

Review—Pathogenesis of Ascites

Place I for increased and D for decreased concerning the physiological factors as they occur with ascites. Give the rationale for each physiological factor. After completing this review, refer to the table for additional information.

_____ Portal obstructions and hypertension.

_____

_____

_____

— — — — — — — — — — — — — —

I Portal hypertension influences ascites, but its presence is associated with liver damage.

_____ Capillary permeability.

_ _ _ _ _ _ _ _ _ _ _ _ _ _ _

I A permeability defect of capillary endothelium contributes to ascites.

_____ Plasma osmotic pressure.

_ _ _ _ _ _ _ _ _ _ _ _ _ _ _

D Due to liver congestion, protein fluid leaves the capillaries and passes into the abdomen.

_____ Retention of sodium and water.

_ _ _ _ _ _ _ _ _ _ _ _ _ _ _

I Increased hormonal response (aldosterone). This decreases urinary output and causes retention of Na and H_2O.

Clinical Considerations

40

Individuals with ascites eventually become refractory in regard to diuretic agents. Frequently, paracentesis (surgical puncture of the abdominal cavity for relieving fluid) is required to relieve symptoms of pressure and/or respiratory distress.

With ascites, there is a tendency for the individual to become refractory to _____.

— — — — — — — — — — — — — —

diuretics

_____

41

Paracentesis means a *_____

_____.

— — — — — — — — — — — — — —

surgical puncture of the abdominal cavity.

_____

42

Paracentesis is generally required to relieve symptoms of _____

_____ or _____.

— — — — — — — — — — — — — —

pressure
respiratory distress

_____

43

Repeated paracenteses result in a great loss of protein, electrolytes, and water.

As a result of repeated paracenteses, what three substances may be lost? _____, _____, and _____.

— — — — — — — — — — — — — —

protein, electrolytes, and water

_____

44

The after-effect of abdominal paracentesis is an antidiuresis of water, with hemodilution of sodium in the blood. Following this, there is a rapid outpouring of fluid into the abdominal cavity with hemoconcentration and a drop in blood volume. This leads to a greater retention of sodium.

Abdominal paracentesis (is/is not) the permanent cure for ascites.

After an abdominal paracentesis, antidiuresis of water results, causing

hemodilution of *_____ .

– – – – – – – – – – – – – –

is not
sodium in the blood

45

After hemodilution of serum sodium following a paracentesis, fluid then

pours into the *_____. Would the blood volume

be increased or decreased? _____. As the result of the

latter, would hemodilution or hemoconcentration be present?

_____ .

The nurse should observe for symptoms of serum sodium (excess/deficit).

– – – – – – – – – – – – – – –

abdominal cavity
decreased
hemoconcentration
excess

46

Repeated paracenteses will cause a great loss in _____,

_____, and _____.

   The removal of a large volume of fluid by paracentesis causes a rapid

shift of fluid from the plasma into the *_____,
Symptoms of circulatory collapse (shock symptoms) should be observed
following the removal of a large volume of abdominal fluid.

Name at least five symptoms of shock.

1. _____

2. _____

3. _____

4. _____

5. _____

– – – – – – – – – – – – – – –

water, protein, and electrolytes
abdominal cavity
1. pallid, cold, clammy skin
2. fast pulse rate
3. apprehension and restlessness
4. fall in blood pressure
5. respirations are shallow and rapid
6. others

Clinical Example

47

Mr. Moore, age 58, was admitted to the medical floor of the hospital complaining of shortness of breath with no chest pain. He had massive ascites with distended veins over the abdomen and 4+ pitted leg edema. Mr. Moore's shortness of breath would most likely be the result of

_____. His records show that his admissions in the past 8 years have been due to cirrhosis of the liver with ascites and peripheral edema. Congestive heart failure frequently accompanies severe liver damage with ascites; therefore, a cardiotonic is given

Mr. Moore was placed on diuretics and a cardiotonic drug (Digoxin).

Diuretics were ordered to increase * _____

_____.

Digoxin was given to * _____

_____.

(Refer to section on edema (CHF) if necessary)

— — — — — — — — — — — — — — — —

ascites
fluid loss via urinary output
strengthen the heart beat and improve circulation of fluid

The laboratory studies of Mr. Moore in Table 32 show how his laboratory results deviate from the norms at the time of his illness.

TABLE 32   Laboratory Studies of Mr. Moore

Laboratory Tests	On Admission	2 Weeks Later	3 Weeks Later	4 Weeks Later
Hematology				
Hemoglobin *(12.9–17.0 grams)	10.4		12.6	
Hematocrit (40–46%)	34		41	
W.B.C. (white blood count) *(5,000–10,000 cu mm)	8,000			
Biochemistry				
B.U.N. (blood urea nitrogen) *(10–15 mg/100 ml)	47	55	132	190
Plasma CO_2 $\frac{*(50-70 \text{ vol }\%)}{*(22-32 \text{ mEq/L})}$	$\frac{44}{20}$	$\frac{48}{22}$		$\frac{29}{13}$
Plasma chloride *(98–107 mEq/L)	105	95	99	91
Plasma sodium *(135–146 mEq/L)	136	124	133	119
Plasma potassium *(3.5–5.3 mEq/L)	4.9	4.6	3.9	4.5
Plasma albumin (3.2–5.6 grams/100 ml)	1.4			

*Laboratory norms from the Wilmington Medical Center, Delaware Division.

48

Mr. Moore's hemoglobin on admission was 10.4 grams, which could be indicative of secondary anemia or (hemodilution/hemoconcentration).

The B.U.N. on his admission was 47 mg/100 ml which would mean

that * _____

_____

_____

_____ .

— — — — — — — — — — — — — —

hemodilution

he is retaining urea—a by-product of protein metabolism which is normally excreted by the kidneys.

49

Two weeks after admission, Mr. Moore's B.U.N. became alarmingly increased. What would this mean in regards to his kidneys? _____

_____

_____ .

— — — — — — — — — — — — —

The inability for the kidneys to excrete the urea, or kidney failure, or a similar response.

50

Mr. Moore's plasma CO_2 on admission was 20 mEq/L which could mean

he was in a mild _____ state.

— — — — — — — — — — — — — —

acidotic

51

Ten days later, Mr. Moore had a paracentesis done. Also, he remained on diuretics. His serum sodium and chloride were low 2 weeks after his admission. This could most likely be caused by the _____

_____ and _____ .

_ _ _ _ _ _ _ _ _ _ _ _ _ _ _

paracentesis and diuretics

52

What would Mr. Moore's plasma albumin indicate, hyperalbuminema or hypoalbuminema? _____ .

_ _ _ _ _ _ _ _ _ _ _ _ _ _

hypoalbuminema

53

Explain where Mr. Moore's albumin could be found? _____

_____ .

_ _ _ _ _ _ _ _ _ _ _ _ _ _

in the peritoneal cavity or abdomen

Clinical Management

54

Mr. Moore received for clinical management:
1. Diuretics
   a. Furosemide 160–200 mg (Stat doses–immediate)
   b. Aldactone (spironolactone) which is an aldosterone antagonist (daily doses)

   Furosemide is a new, potent diuretic which acts on the proximal and distal tubules and ascending limb of Henle's loop. If given in excessive amounts, Furosemide can lead to a profound diuresis with water and electrolyte depletion.

Aldactone (spironolactone) inhibits the production of aldosterone (the hormone which causes sodium and water retention and potassium excretion). Therefore, Aldactone promotes sodium and water excretion and inhibits potassium excretion.

Thiazide diuretics are not indicated since they cause potassium excretion. Hypokalemia can cause hepatic coma or liver failure. Low serum potassium has a tendency to increase ammonium accumulation which precipitates hepatic toxicity.

2. Digoxin—0.25 mg daily
3. Low sodium diet—1.5 grams
4. Limited fluid intake
5. Daily weights

Mr. Moore received Furosemide by _____ doses. It is a potent diuretic which acts on what areas of the kidneys? _____

_____.

Large and continuous doses of Furosemide can lead to a profound diuresis causing depletion of _____ and _____.

Aldactone promotes excretion of _____ and _____ and inhibits excretion of _____.

Thiazide diuretics are contraindicated because *_____

_____ which could lead to *_____.

Digoxin was given daily to *_____ and

_____.

Low sodium diet and limited fluid intake would help to decrease the body's _____ and _____.

– – – – – – – – – – – – – – –

stat
proximal and distal tubules and the loop of Henle
water and electrolytes
sodium and water, potassium
they cause potassium excretion which could lead to hepatic coma
strengthen heart beat and improve circulation
sodium and water

55

Mr. Moore did not respond well to Furosemide and Aldactone. Paracentesis was then indicated when his ascites became refractory to _____

_____ .

Paracentesis was done on the tenth day after admission and during the third week.

Following the removal of a large volume of fluid by paracentesis, the

nurse should observe for symptoms of _____

_____ .

– – – – – – – – – – – – – – – –

diuretics, circulatory collapse (shock)

56

On several occasions, Mr. Moore received 300 cc of 3% saline (hypertonic) intravenously over 3 hours. Fifteen minutes after the fluid was started I. V., Furosemide 200 mg was given. The purpose of a hypertonic solution and a potent diuretic would be to draw the fluid from

the *_____ into the *_____

_____ for urinary excretion.

Once the major symptoms have been relieved through diuretics, paracentesis, and saline I. V. administration, the physician will be able to treat the basic problem, which is cirrhosis of the liver.

– – – – – – – – – – – – – – – –

abdominal cavity
intravascular spaces

Review—Cirrhosis of the Liver

Q. 1

Liver disease is frequently associated with retention of _____ and

_____ caused by increased portal pressure and increased secretion

of _____ .

_ _ _ _ _ _ _ _ _ _ _ _ _ _ _

sodium and water
aldosterone

_____

Q. 2

Hyponatremia occurs in liver diseases in spite of the presence of an excess of total body sodium because water is *_____

_____ .

_ _ _ _ _ _ _ _ _ _ _ _ _ _

retained in excess to sodium

_____

Q. 3

Other contributing causes of hyponatremia are:

1. _____

_____

2. _____

_____

3. _____

_ _ _ _ _ _ _ _ _ _ _ _ _ _

1. sodium moves into the intracellular spaces
2. prolonged use of potent diuretics
3. low sodium diet

_____

Q. 4
Potassium loss occurs due to:

1. _____

2. _____

3. _____

— — — — — — — — — — — — —

1. increased aldosterone secretion
2. malnutrition due to anorexia
3. diuresis
4. dehydration

Q. 5
Ascites is *_____

_____ .

— — — — — — — — — — — — — —

an accumulation of fluid in the peritoneal cavity (or a similar answer)

Q. 6
Portal hypertension influences fluid accumulation in the abdomen, but

the fundamental cause of ascites lies in *_____

_____ .

— — — — — — — — — — — — — —

damage to the cellular structure of the liver

Q. 7
With ascites, there is a capillary permeability (increase/decrease) and a
plasma osmotic pressure (increase/decrease).

— — — — — — — — — — — — — —

increase
decrease

Q. 8
Individuals with ascites may eventually become refractory to

_____.

_ _ _ _ _ _ _ _ _ _ _ _ _ _

diuretics

Q. 9
Paracentesis is a surgical puncture of the *_____

_____. It is required in order to relieve symptoms of _____

_____.

_ _ _ _ _ _ _ _ _ _ _ _ _ _

abdominal cavity
pressure or respiratory distress

Q. 10
Repeated paracenteses result in a great loss of _____,

_____, and _____.

_ _ _ _ _ _ _ _ _ _ _ _ _ _

protein, electrolytes, and water

Q. 11
Following an abdominal paracentesis, antidiuresis of water results; then

there is a rapid outpouring of fluid into *_____

_____ resulting in hemoconcentration and a drop in

*_____.

_ _ _ _ _ _ _ _ _ _ _ _ _ _

the abdominal cavity
blood volume

Q. 12

Mr. Moore received diuretics as part of his clinical management. Furo-semide is a _____ and when used in large and continuous doses can lead to profound diuresis with depletion of _____ _____ and _____.

– – – – – – – – – – – – – – –

potent diuretic
water and electrolytes

Q. 13

Aldactone promotes excretion of _____ and _____ and inhibits excretion of _____.

– – – – – – – – – – – – – – –

sodium and water
potassium

Q. 14

Digoxin was given to Mr. Moore to *_____ _____ and *_____.

– – – – – – – – – – – – – – –

strengthen his heart beat and improve circulation

Q. 15

What condition will frequently accompany cirrhosis of the liver with ascites? _____.

– – – – – – – – – – – – – – –

congestive heart failure; also renal failure.

RENAL FAILURE AND PERITONEAL DIALYSIS

Physiological Factors

57

Renal failure is when the kidneys have lost their ability to excrete waste products of the body's metabolism.

If the kidneys can no longer excrete waste products, this would indicate *_____.

_ _ _ _ _ _ _ _ _ _ _ _ _ _ _ _

renal failure

58

Acute renal failure is when the kidneys excrete small volumes of urine, less than 400 ml daily, following severe damage to the kidney tissue.

Chronic renal failure usually develops over many months or years, causing a progressive loss of kidney tissue and kidney function.

Renal failure indicates that the kidneys can no longer *_____

_____.

_ _ _ _ _ _ _ _ _ _ _ _ _ _ _

excrete waste products of the body's metabolism

59

Place ARF for acute renal failure and CRF for chronic renal failure concerning the following statements pertaining to renal dysfunction.

_____ a. It causes a progressive loss of kidney tissue and kidney function.

_____ b. It occurs as a result of severe damage to the kidney tissue, and less than 400 ml of urine is excreted daily.

_ _ _ _ _ _ _ _ _ _ _ _ _ _ _

a. CRF
b. ARF

60

Acute renal failure is when the kidneys excrete small volumes of urine,

less than _____, following *_____

_____.

Chronic renal failure usually develops over many months or years, caus-

ing *_____.

– – – – – – – – – – – – – – –

400 ml daily,
severe damage to the kidney tissue
progressive loss of kidney tissue and kidney function

61

Acute renal failure can be divided into two phases: oliguric phase and
diuretic phase.

 Oliguric phase is when the urinary volume is less than 400 ml a day
and diuretic phase is when the urinary volume is approaching a liter
(1000 ml) a day.

 When the urinary volume is less than 400 ml with ARF, the individual

is in the _____ phase and when it is approaching a liter per

day, the individual is in the _____ phase.

– – – – – – – – – – – – – – –

oliguric
diuretic

 Table 33 outlines the clinical features, causes, and specific fluid and
electrolyte changes associated with acute and chronic renal failure.
Study this table carefully so you can distinguish the characteristics of
acute and chronic renal failure. You may need to use the glossary for
unknown words. Refer to this table as needed.

TABLE 33 Characteristics of Acute and Chronic Renal Failure

Classification	Acute Renal Failure	Chronic Renal Failure
Clinical features	Weakness Nausea and vomiting Muscular twitching Pruritis B.U.N. increased Urinary output less than 400 ml	Lassitude Loss of weight Hypertension Oliguric to anuric Later symptoms—skin and GI bleeding, muscular twitching, and uremic convulsions
Causes	Burns, severe wounds, crushing injuries, hemolytic blood transfusion reaction, accidental infusion of distilled water Nephrotoxic substances, i.e., bichloride of mercury, carbon tetrachloride, lead Severe infections Severe fluid volume deficit Heart failure	Glomerulonephritis, pyelonephritis, polycystic kidneys, and hydronephrotic damage Essential hypertension Urinary obstruction
Fluid and Electrolyte Changes		
Overhydration	Fluid volume excess occurs from excessive administration of fluids, orally or parenterally, during the oliguric phase. Symptoms of overhydration are neck vein distention, edema, bounding pulse, and shortness of breath would be present	Fluid volume excess occurs from excessive administration of fluids, orally or parenterally. Symptoms of overhydration are the same as for ARF.

Metabolic acidosis	Metabolic acidosis results from: 1. Inability to excrete acid metabolites. 2. Decreased food intake causing increased utilization of body fats and the accumulation of ketonic acids.	Metabolic acidosis results from: 1. Inability to excrete acid metabolites. 2. Inability of the tubules to form ammonia.
	Respiratory compensatory mechanism will partially relieve the acidotic condition by exhaling CO_2.	Respiratory compensatory mechanism will partially relieve the acidotic condition by exhaling CO_2.
Potassium	Potassium excess occurs when: 1. The urinary output is reduced. 2. There is massive tissue destruction causing potassium to leave the cells, thus the serum potassium is increased. The kidneys normally excrete 80% of the daily potassium loss.	Potassium varies: Potassium excess is associated with severe oliguria and usually does not occur until late in chronic renal failure. Potassium deficit could occur with chronic nephritis having polyuria (excessive amount of urine).
Sodium deficit	Sodium deficit results from: 1. Administering excessive amounts of water that dilute serum sodium. 2. The shift of sodium into the cells, especially if acidosis is present. 3. Vomiting and diarrhea.	Sodium deficit results from: 1. A low sodium diet. 2. Vomiting and diarrhea. Otherwise, the serum sodium remains normal.

Calcium deficit	Calcium deficit may be related to an increase in serum phosphorus, which is a retained constituent of metabolic acids. An increase in one will cause a decrease in the other.	When metabolic acidosis is present, tetany symptoms are not common though muscle twitching may occur.
	Calcium deficit frequently does not present symptoms of tetany because the acidotic condition favors calcium ionization. If the individual becomes alkalotic due to alkaline fluids, then tetany symptoms are present because of a lack of calcium ionization.	Same as for ARF.
	Calcium deficit will enhance the toxic effect of potassium on the heart since they have antagonistic actions on heart muscle when in deficit.	
Other changes	Anemia: Etiology is unknown though it is felt to be due to the level of azotemia (excessive quantities of nitrogenous waste products in the blood). It is felt there is an increase in blood destruction.	Anemia: Its severity is related to the degree of azotemia. It is felt there is an increase in blood destruction.
	Uremia: If oliguria is not relieved, then uremia progresses.	If severe, it can contribute to: 1. the development of CHF and acidosis. 2. G.I. bleeding because of the failure of the platelet factor. 3. pericarditis. 4. pleurisy.

62

Indicate which of the following causes contribute to acute and chronic renal failure by using <u>ARF</u> for acute renal failure and <u>CRF</u> for chronic renal failure.

_____ Burns

_____ Crushing injuries or severe wounds

_____ Glomerulonephritis

_____ Pyelonephritis

_____ Polycystic kidneys

_____ Urinary obstruction

_____ Essential hypertension

_____ Toxins from bichloride of mercury, carbon tetrachloride, lead

_____ Severe infections

_____ Hemolytic blood transfusion reaction

_____ Severe fluid volume deficit

— — — — — — — — — — — — — — —

ARF, ARF, CRF, CRF, CRF, CRF, CRF, ARF, ARF, ARF, ARF

63

In acute renal failure, potassium excess occurs when:

a. _____

b. _____

The kidneys normally excrete _____% of the daily potassium loss.

— — — — — — — — — — — — — — —

a. the urinary output is reduced.
b. there is massive tissue destruction.
80%

64

In chronic renal failure, high potassium levels do not usually occur until later on in its progress and when what is present? _____

_____.

– – – – – – – – – – – – – – –

severe oliguria

65

With acute and chronic renal failure, overhydration occurs from

*_____

_____.

Symptoms of overhydration which the nurse would observe are:

a. _____

b. _____

c. _____

d. _____

– – – – – – – – – – – – – –

an excessive administration of fluids—orally and parenterally

a. neck vein distention   c. bounding pulse

b. edema   d. shortness of breath

66

With acute renal failure, metabolic acidosis results from the inability to excrete *_____ and from the accumulation of *_____.

  With chronic renal failure, metabolic acidosis results from the inability to excrete *_____ and from the inability of the tubules to form _____.

– – – – – – – – – – – – – –

acid metabolites; ketonic acids
acid metabolites; ammonia

67

Respiratory compensatory mechanism will partially relieve the metabolic condition by *_____.

— — — — — — — — — — — — — —

exhaling CO_2

68

With acute renal failure, sodium deficit can result from:

a. _____

b. _____

c. _____

With chronic renal failure, sodium deficit can result from:

a. _____

b. _____

— — — — — — — — — — — — — —

a. administering excessive amounts of water
b. the shift of sodium into cells
c. vomiting and diarrhea
a. a low sodium diet
b. vomiting and diarrhea

69

Calcium deficit may be related to an increase in serum _____.

This element is a retained constituent of _____.

  Calcium deficit frequently does not present symptoms of tetany because of the _____ condition.

  Calcium deficit will enhance the toxic effect of _____
on the heart.

— — — — — — — — — — — — — —

phosphorus   acidotic
metabolic acids   potassium

70

Anemia is thought to be due to the level of _____ .

Azotemia is the *_____

_____

_____ .

— — — — — — — — — — — — — —

azotemia
excessive quantities of nitrogenous waste products (compounds) in the blood

71

Bleeding is thought to be related to the failure of the *_____

_____ .

Other conditions that might accompany chronic renal failure are

_____ , _____ , and _____

_____ .

— — — — — — — — — — — — — —

platelet factor
pericarditis, pleurisy, and congestive heart failure

Review of Table 33

For the following fluid and electrolyte changes, place <u>ARF</u> for acute renal failure, <u>CRF</u> for chronic renal failure, and <u>O</u> for neither ARF or CRF as they apply to these changes.

_____   a. Potassium excess

_____   b. Potassium deficit

_____   c. Overhydration—fluid volume excess

_____   d. Dehydration—fluid volume deficit

_____   e. Metabolic acidosis

_____   f. Metabolic alkalosis

_____   g. Sodium excess

_____   h. Sodium deficit

_____   i. Calcium excess

_____   j. Calcium deficit

– – – – – – – – – – – – – – –

a. ARF, CRF
b. CRF with chronic nephritis
c. ARF, CRF
d. ARF—as a cause
e. ARF, CRF
f. O
g. O
h. ARF, CRF
i. O
j. ARF, CRF

Clinical Considerations

72

The nurse caring for individuals with kidney dysfunction should observe for symptoms of potassium excess, fluid retention, metabolic acidosis, sodium deficit, calcium deficit, and for signs and symptoms of anemia.

While caring for patients having renal dysfunction, the nurse would be observing for signs and symptoms of:

a. metabolic _____

b. sodium _____

c. fluid _____

d. potassium _____

e. calcium _____

_ _ _ _ _ _ _ _ _ _ _ _ _ _ _

a. acidosis
b. deficit
c. retention
d. excess
e. deficit

Dialysis

73

To maintain the life of patients afflicted with renal failure, either acute or chronic, various types of dialysis can be initiated. <u>Dialysis</u> is a method used to eliminate body waste products and unneeded blood constituents. The products frequently eliminated would include: protein catabolism, e.g., urea, salts (excess electrolytes), acid metabolites, toxic substances, and water.

Define dialysis. _____

_____

_____.

 Dialysis can be initiated for patients with either _____

or _____ renal failure.

– – – – – – – – – – – – – – – –

A method used to eliminate body waste products and unneeded blood constituents.

acute or chronic

74

The two types of dialysis frequently employed are <u>hemodialysis</u> or the artificial kidney, and <u>peritoneal dialysis</u>.

Emphasis will be placed on peritoneal dialysis since this type is most commonly used in many hospitals. Though hemodialysis is most effective, it is still considered a costly method of dialysis, and not all hospitals have a renal unit for this type of dialysis.

Name at least five types of products that might be eliminated through the process of dialysis.

1. _____

2. _____

3. _____

4. _____

5. _____

— — — — — — — — — — — — — — —

1. protein catabolism, e.g., urea
2. electrolytes or salts
3. acid metabolites
4. toxic substances
5. water

75

The peritoneal lining that surrounds the abdominal cavity can be used as a dialyzing membrane. What type of membrane would you say the

peritoneal lining is? _____.

— — — — — — — — — — — — — — —

selectively permeable

76

The fluid which enters the peritoneal cavity via catheter resembles plasma in electrolyte concentration and tonicity, and is in osmotic and chemical equilibrium with normal blood and interstitial fluid. Alteration in fluid volume is accomplished through the use of various concentrations of glucose or dextrose in the dialyzing solution.

What is the name of the membrane used for this type of dialysis?

_____ .

Name the substance that will alter the fluid volume.

_____ .

– – – – – – – – – – – – – – – –

peritoneal membrane
glucose or dextrose

77

The fluid used for peritoneal dialysis resembles plasma in _____

_____ and _____ .

– – – – – – – – – – – – – – – –

electrolyte concentration and tonicity

78

The solution must be at least slightly hypertonic (hyperosmotic) to prevent its absorption by the plasma which could lead to fluid volume excess (overhydration).

Dextrose, in different concentrations, is used to render the solution hypertonic.

If the dialyzing solution (peritoneal dialysate) is not slightly hypertonic, the plasma will absorb the solution and *_____

_____ will develop.

The ingredient used in dialysis to achieve the varying degrees of solution concentration is _____ .

– – – – – – – – – – – – – – – –

fluid volume excess or overhydration   dextrose or glucose

79

The two strengths frequently used are the 1–1/2% or 7% of <u>dextrose</u> dialysate. In either strength, the chemical composition is equivalent to normal serum concentration. The ingredient responsible for its degree

of concentration is _____ .

— — — — — — — — — — — — — — — —

dextrose or glucose

80

The chemical composition of dialysate is as follows:

Na — 137 mEq/L
Cl — 102 mEq/L
Ca —   4 mEq/L
Mg — 1.5 mEq/L
HCO$_3$ — 43 mEq/L (as lactate)
K   — None is added (appropriate amounts will be added according to the patient's serum potassium)

The chemical composition of the peritoneal dialysis solution is _____ .

_____ , _____ , _____ ,

and _____ .

   Potassium is added to the solution in appropriate amounts according

to.*_____

_____ .

— — — — — — — — — — — — — — —

sodium chloride, calcium, magnesium, and lactate (HCO$_3$)
the patient's serum potassium

81

For a patient with a normal serum potassium, KCl (4 mEq/L) will be added to the dialysis in order not to remove potassium from the patient.

Do you think KCl would be added to the dialysis for patients with

hyperkalemia? _____.

Explain. _____.

– – – – – – – – – – – – – – –

no

It would increase the already high serum potassium.

82

Heparin, an anticoagulant, is sometimes added for the prevention of fibrin and clots from obstructing the peritoneal dialysis catheter. The

purpose for using heparin is to prevent *_____

_____.

– – – – – – – – – – – – – – –

fibrin and clot formation

Table 34 gives the names of the dextrose dialysate solutions commercially produced by three manufacturers of parenteral fluids and gives the uses according to the percentage of the solution. These solutions are prepared in two strength of dextrose, 1½% and 7%. When 4½% dextrose dialysate solution is used, a bottle of 1½% and one of 7% is administered simultaneously through the abdominal catheter by using Y tubing.

Study this table carefully. You will not be expected to memorize the manufacturer's name of the solution, but you will be required to know the three strengths of the solutions and when they are indicated for use. Refer to this table as needed.

TABLE 34   Various Peritoneal Dialysis Solutions and Their Uses

Solutions and Their Uses	Slightly Hypertonic (Hyperosmotic)	Moderately Hypertonic (Hyperosmotic)	Highly Hypertonic (hyperosmotic)
Uses	When edema is not present, but electrolytes need adjustment.	When moderate edema is present and electrolytes need adjusting.	When severe edema is present and rapid removal is indicated.
Abbott's Impersol with dextrose	1–1/2%	4–1/2% (combination of a liter of 1–1/2% and a liter of 7%)	7%
Baxter's Dianeal with dextrose	1–1/2%	4–1/2% (combination of a liter of 1–1/2% and a liter of 7%)	7%
Cutter's Peridial with dextrose	1–1/2%	4–1/2% (combination of a liter of 1–1/2% and a liter of 7%)	7%

83

Slightly hypertonic, _____%, peritoneal dialysis solution is used
when *_____

_____

_____.

   Moderately hypertonic, _____%, peritoneal dialysis is used
when *_____

_____

_____.

   Highly hypertonic, _____%, peritoneal dialysis solution is used
when *_____

_____

_____.

— — — — — — — — — — — — — —

1–1/2%
edema is not present, but electrolytes need adjustment
4–1/2%
moderate edema is present and electrolytes need adjustment
7%
severe edema is present and rapid fluid removal is indicated

84

Positive fluid balance indicates fluid retained by the patient and negative fluid balance indicates fluid withdrawn from the patient.

If a patient receives 1000 ml of 1½% dextrose dialysate and 500 ml was returned or siphoned off, then the patient would have retained 500 ml of the solution. Would the patient be in positive or negative balance? _____.

Could dehydration or overhydration occur due to this fluid imbalance?

_____.

If a patient receives 1000 ml of 7% dextrose dialysate and 2000 ml was returned or siphoned off, how much fluid was withdrawn in excess to the amount administered? _____. Would the patient be in positive or negative fluid balance? _____.

Could dehydration or overhydration occur due to this fluid imbalance?

_____.

– – – – – – – – – – – – – – – –

positive fluid balance
overhydration
1000 ml
negative fluid balance
dehydration

85

A positive balance of 500 ml or negative balance of 1000 ml should be brought to the physician's attention.

Define positive fluid balance. _____

_____.

Define negative fluid balance. _____

_____.

– – – – – – – – – – – – – – – –

Fluid retained by the patient
Fluid withdrawn from the patient

86

A positive balance of _____ or a negative balance of

_____ should be brought to the physician's attention.

— — — — — — — — — — — — — — —

500 ml
1000 ml

87

The nurse should be observing for symptoms of fluid volume excess
(overhydration) and symptoms of fluid volume deficit (dehydration)
while the patient is receiving peritoneal dialysis.

  Give four symptoms of overhydration.

_____ , _____ , _____

_____ , and _____ .

  Give at least five symptoms of marked dehydration.

_____ , _____ ,

_____ , and _____ .

— — — — — — — — — — — — — — —

neck vein distention, edema, bounding pulse, and shortness of breath.
marked thirst, dry mucous membrane, wrinkled skin, tachycardia, and
a low elevated temperature

Clinical Example

Mrs. Grady, age 21, had not been able to void for the last 24 hours nor had she any sensation of a need to void when she was admitted to the intensive care unit. There was noted fullness in her face, back, and abdomen, but no indication of bladder distention. She had +1 pitted edema of the ankles. She was also having mild dyspnea. Before admission, she had noted a decrease in urine output for the last 4 days. She stated she had been having diarrhea.

This was Mrs. Grady's fourth admission to this hospital. She had a history of hypertension for 7 years. The diagnoses were:
1. chronic pyelonephritis with severe nephrosclerosis.
2. complete renal shutdown
3. hypertensive cardiovascular disease secondary to rheumatic heart disease
4. congestive heart failure
5. anemia—secondary to chronic renal disease

88

The fullness in her face, back, and abdomen was contributed to (edema/dehydration).

The diagnosis of pyelonephritis with severe nephrosclerosis was an indication of (acute/chronic) renal failure.

On admission, her B.U.N. was 78 mg/100 ml and creatinine 4.0 mg/100 ml, which indicated that there was an accumulation of *_____
_____.

– – – – – – – – – – – – – – – –

edema
chronic
nitrogenous waste products in the blood (azotemia)

Table 35 gives the laboratory studies for Mrs. Grady and show how her results differ from the norms at the time of her illness.

TABLE 35 Laboratory Studies for Mrs. Grady

Laboratory Tests*	On Admission	1st Week	4th Week (1st Day)	4th Week (2nd Day)	4th Week (3rd Day)	5th Week (1st Day)	5th Week (2nd Day)
Hematology							
Hemoglobin (12.9–17.0 grams)	6.1		8.0	6.8			
Hematocrit (40–46%)	19		26	23			
W.B.C. (white blood count) (5,000–10,000 cu mm)	4,500		13,600				
Biochemistry							
B.U.N. (blood urea nitrogen) (10–15 mg/100 ml)	78	54	80	36	31	25	33
Creatinine (1–2 mg/100 ml)	4	–	–	–			
Plasma CO_2 (22–32 mEq/L)	13	20	9	24	26		
Plasma chloride (98–107 mEq/L)	92	95	85	92	94		
Plasma sodium (135–146 mEq/L)	121	132	132	136	133		
Plasma potassium (3.5–5.3 mEq/L)	5.6	4.5	4.2	2.8	3.1	4.6	
Plasma calcium (9–11 mg/100 ml)	8.8						7.0

*Laboratory norms from Wilmington Medical Center, Delaware Division.

89

On admission, Mrs. Grady's hemoglobin of 6.1 grams and hematocrit of

19% indicated _____ which was secondary to * _____

_____ .

– – – – – – – – – – – – – – – –

anemia
chronic renal disease

90

Mrs. Grady's CO_2 combining power of 13 mEq/L on admission indicated

she was in a state of * _____ .
    Her sodium deficit could have occurred from excessive fluid intake
diluting the serum sodium, the sodium shift into the cells, and/or

_____ .

    Her chloride deficit could have occurred as a result of _____

_____ and _____ .
Her potassium excess could be the result of:

1. _____ .

2. _____ .

    _____ .

– – – – – – – – – – – – – – – –

metabolic acidosis
diarrhea
diarrhea and sodium deficit
1. reduced urinary output.
2. tissue destruction, or as a result of the sodium shift.

91

Which of the following changes of Mrs. Grady indicated that she was in chronic renal failure.

_____ a. Hemoglobin 6.1 grams and hematocrit 19%

_____ b. B.U.N. 78 mg/100 ml and creatinine 4 mg/100 ml

_____ c. Plasma CO_2 13 mEq/L

_____ d. Plasma Cl 92 mEq/L

_____ e. Plasma Na 121 mEq/L

_____ f. Plasma K 5.6 mEq/L

_____ g. Plasma Ca 8.8 mg/100 ml

_ _ _ _ _ _ _ _ _ _ _ _ _ _ _

a. X, b. X, c. X, d. X, e. X, f. X, g. X

Clinical Management

The clinical management pertains to Mrs. Grady and can pertain to others receiving peritoneal dialysis.

92

Peritoneal dialysis was started on Mrs. Grady the day she was admitted. At first, she received a total of 5 liters of 7% dialysate and then was switched to 1½% solution. Her dialysis cycle was one liter per hour (10 minutes for the solution to run in, 30 minutes to remain in the peritoneal cavity, and 20 minutes to siphon out).

The total fluid withdrawn (output) from her body using 5 liters of 7% dialyzing solution was 7200 ml. How much fluid was withdrawn in excess to the 5000 ml administered? _____.

Would Mrs. Grady be in positive or negative fluid balance _____

_____.

_ _ _ _ _ _ _ _ _ _ _ _ _ _ _

2200 ml   negative fluid balance

93

When using a high hypertonic solution, water shifting from the intravascular space and interstitial spaces into the peritoneal cavity may be so great that circulatory collapse or _____ could occur.

– – – – – – – – – – – – – – –

shock

94

The nurse should observe for respiratory embarrassment or distress due to the diaphragm pushed upward from using 2 liters of solutions.

The symptom of respiratory embarrassment would be _____.

– – – – – – – – – – – – – – –

dyspnea

95

From using excessive amounts of 7% dialysate, water will shift from the

* _____

into the * _____.

This water shift could be so great that _____ could occur.

– – – – – – – – – – – – – – –

extracellular spaces (intravascular and interstitial spaces) into the peritoneal cavity
shock

96

As a result of 2200 ml fluid loss, Mrs. Grady started to dehydrate so

fluids were given to maintain *_____ .

When using 7% dialysate, (dehydration/water intoxication) can occur

and when using a lower glucose concentration, *_____

_____ can occur if there is a rapid removal of solute by
dialysis.

─ ─ ─ ─ ─ ─ ─ ─ ─ ─ ─ ─ ─ ─ ─ ─

fluid balance or homeostasis
dehydration
water intoxication

97

Mrs. Grady's diet was restricted of potassium, protein, and fluids. The
foods restricted from her diet would be: meats, fresh fruits, fruit juices,
legumes, nuts, milk, tea, and coffee (decaffinated).

Protein foods will increase the blood urea and the body's potassium
and water.

The foods that should be restricted from Mrs. Grady's diet are:

(   ) a. bread

(   ) b. meats

(   ) c. fresh fruits

(   ) d. fruit juices

(   ) e. nuts

(   ) f. milk

(   ) g. tea and coffee

─ ─ ─ ─ ─ ─ ─ ─ ─ ─ ─ ─ ─ ─

a. −, b. X, c. X, d. X, e. X, f. X, g. X

98

Protein foods will increase _____ in the blood and also the body's

_____ and _____ .

– – – – – – – – – – – – – – –

urea
potassium and water

99

Fluid intake should replace only fluid losses, e.g., from urinary output, vomiting, and diarrhea. Also, insensible perspiration accounts for 500–600 ml of fluid loss. If an excess amount of fluid is retained, then the fluid loss following treatment may be greater than the fluid intake.

Accurate daily weights can also determine gain or loss of body fluid.

Mrs. Grady's fluid intake was calculated according to her _____

_____ .

Insensible perspiration would account for _____ ml of fluid loss.

Mrs. Grady was weighed daily, which aided the physician in determin-

ing *_____

_____ .

– – – – – – – – – – – – – –

fluid losses.
500–600 ml
fluid intake and body fluid retainment

100

Frequently, sodium deficit is treated by limiting the water intake. Administering hypertonic solution of sodium chloride for correction of sodium deficit could result in fluid volume excess with congestive heart failure and pulmonary edema, especially if water intoxication is occurring.

To treat sodium deficit, water intake would be (limited/increased).

Administering hypertonic solution of sodium chloride when water

intoxication is present could lead to *_____

_____.

_ _ _ _ _ _ _ _ _ _ _ _ _ _ _

limited

congestive heart failure and pulmonary edema or fluid volume excess

101

During the fourth week, Mrs. Grady again received peritoneal dialysis. After the second day of the fourth week, when Mrs. Grady's serum potassium was 2.8 mEq/L, potassium chloride, 8 mEq/L, was added to each dialysis cycle.

This was necessary because Mrs. Grady's serum potassium indicated a

_____ state, which was most likely the result of dialysis

due to insufficient _____ replacement.

During dialysis, serum electrolytes should be checked frequently. Can

you explain why?_____

_____

_____.

_ _ _ _ _ _ _ _ _ _ _ _ _ _ _

hypokalemic

potassium

There could be a rapid loss of electrolytes through dialysis, or a rapid gain from electrolytes added to dialyzing solution.

102

During the fifth week, Mrs. Grady's serum potassium became 7.0 mEq/L.

Cation exchange resins will help to excrete potassium from the intestines. The resins may be given orally or as a retention enema.

Kayexalate (sodium polystyrene) is a cation exchange resin which when given with mannitol or sorbitol will induce an "osmotic diarrhea."

When using Kayexalate, for every potassium ion excreted, a _____ ion is absorbed from the intestine.

_ _ _ _ _ _ _ _ _ _ _ _ _

sodium

103

The resin helpful in excreting potassium from the intestine is _____

_____.

_ _ _ _ _ _ _ _ _ _ _ _ _

Kayexalate or sodium polystyrene

104

Symptoms of hyperkalemia are:

( ) a. abdominal distention
( ) b. abdominal cramps
( ) c. tachycardia and later bradycardia
( ) d. dizziness
( ) e. numbness or tingling

_ _ _ _ _ _ _ _ _ _ _ _ _

a. —
b. X
c. X
d. —
e. X

Review—Renal Failure and Peritoneal Dialysis

Q. 1

Renal failure indicates that the kidneys have lost their ability to *_____
_____.

— — — — — — — — — — — — — — — —

excrete waste products of the body's metabolism
_____

Q. 2

Acute renal failure is when the kidneys excrete small volumes of urine,

less than _____, following *_____
_____.

— — — — — — — — — — — — — — — —

400 ml daily,
severe damage to the kidney tissue
_____

Q. 3

Chronic renal failure usually develops over many months or years, caus-

ing a progressive loss of *_____.

— — — — — — — — — — — — — — — —

kidney tissue and kidney function
_____

Q. 4

The kidneys normally excrete _____% of the daily potassium loss.

— — — — — — — — — — — — — — — —

80%

_____

Q. 5
Potassium excess occurs in acute and chronic renal failure when:

1. _____

2. _____

_ _ _ _ _ _ _ _ _ _ _ _ _ _ _

1. urinary output is reduced
2. there is massive tissue destruction

Q. 6
With acute and chronic renal failure overhydration occurs from *_____

_____

_____ .

_ _ _ _ _ _ _ _ _ _ _ _ _

excessive administration of fluids—orally and parenterally

Q. 7
Metabolic acidosis can occur in acute and chronic renal failure from:

1. _____

_____

2. _____

_____

_____

3. _____

_ _ _ _ _ _ _ _ _ _ _ _ _ _

1. inability to excrete acid metabolites
2. accumulation of ketonic acids from utilization of body fats
3. inability of the tubules to form ammonia

Q. 8
Sodium deficit in acute and chronic renal failure can result from:

1. _____

   _____

2. _____

   _____

3. _____

4. _____

— — — — — — — — — — — — — — —

1. administering excessive amounts of water
2. the shift of sodium into the cells
3. vomiting and/or diarrhea
4. a low sodium diet

Q. 9
Calcium deficit may be related to an increase in serum _____

_____. This element is a retained constituent of * _____

_____.

— — — — — — — — — — — — — — —

phosphorus
metabolic acids

Q. 10
Calcium deficit will enhance the toxic effect of the ion _____

_____ on the heart.

— — — — — — — — — — — — —

potassium

Q. 11

While caring for patients having renal dysfunction, the nurse would be observing for signs and symptoms of:

potassium _____

fluid _____

metabolic _____

sodium _____

calcium _____

– – – – – – – – – – – – – – –

potassium excess
fluid retention
metabolic acidosis
sodium deficit
calcium deficit

Q. 12

The fluid used for peritoneal dialysis resembles plasma in * _____

_____ and _____ .

– – – – – – – – – – – – – – –

electrolyte concentration and tonicity

Q. 13

Solutions must be at least slightly _____ to plasma

to prevent _____ and _____

_____ .

– – – – – – – – – – – – – – –

hypertonic
its absorption and development of overhydration

Q. 14

The ingredient used in dialysis to achieve the varying degrees of solution

concentration is _____ .

_ _ _ _ _ _ _ _ _ _ _ _ _ _ _ _

dextrose or glucose

Q. 15

Potassium is added to the solution in appropriate amounts according to

* _____

_____ .

_ _ _ _ _ _ _ _ _ _ _ _ _ _ _

the patient's serum potassium

Q. 16

With peritoneal dialysis, a positive fluid balance indicates * _____

_____ and a negative fluid balance

indicates * _____

_____ .

_ _ _ _ _ _ _ _ _ _ _ _ _ _ _

fluid retained by the patient
fluid withdrawn from the patient

Q. 17

A positive balance of over _____ or negative balance of over

_____ should be brought to the physician's attention.

_ _ _ _ _ _ _ _ _ _ _ _ _ _ _

500 ml
1000 ml

Q. 18

When using a high hypertonic solution, circulatory collapse or

_____ could occur.

– – – – – – – – – – – – – –

shock

Q. 19

Respiratory embarrassment or distress can occur as the result of perito-

neal dialysis because of *_____

_____.

– – – – – – – – – – – – – –

the fluid pushing the diaphragm upward

Q. 20

Mrs. Grady's diet was restricted of potassium, protein, and fluids. The

foods restricted from her diet would be _____ , _____

_____ , _____ , _____ ,

_____ , _____ , and _____ .

– – – – – – – – – – – – – –

meats, fresh fruits, fruit juices, legumes, nuts, milk, tea, and coffee

Q. 21

Protein foods will increase _____ in the blood and increase the

body's _____ and _____ .

– – – – – – – – – – – – – –

urea
potassium and water

Q. 22

The cation exchange resin, Kayexalate, is helpful in the excretion of

_____ from the (stomach/intestines).

— — — — — — — — — — — — — — —

potassium
intestines

BURNS

Physiological Factors

105

Following a burn, there is a shift of fluid and electrolytes from the plasma to the interstitial spaces of the burned areas. This would result in a(an) (increase/decrease) of circulating plasma volume.

— — — — — — — — — — — — — —

decrease

106

With a decrease in circulating plasma volume, the following can occur:
1. shock
2. renal shutdown
3. accumulation of serum potassium

More water than protein leaves the intravascular space, increasing the concentration of circulating plasma protein. With the increase in the plasma protein, there will be a(an) (increased/decreased) osmotic pressure resulting in fluid moving from the cells to the plasma and then into

the *_____.

This will result in (water intoxication/dehydration).

— — — — — — — — — — — — — —

increased
interstitial space or fluid
dehydration

107

Burns will cause (more/less) water than protein to leave the intravascular space.

The following will then occur:

( ) a. water intoxication
( ) b. dehydration
( ) c. increased concentration of plasma protein
( ) d. decreased concentration of plasma protein
( ) e. increased plasma osmotic pressure
( ) f. decreased plasma osmotic pressure

— — — — — — — — — — — — — —

more
a. —, b. X, c. X, d. —, e. X, f. —

108

With a resultant lowered blood volume, the following can occur:

1. _____

2. _____

3. _____

— — — — — — — — — — — — — —

1. shock
2. renal shutdown
3. accumulation of serum potassium

109

Erythrocytes or red blood cells are destroyed due to hemolysis (destruction of red blood cells) and bleeding occurring at the burned area. The free hemoglobin released from the red blood cells may produce renal damage.

Erythrocytes are destroyed as a result of:

( ) a. bleeding into the burned area
( ) b. increased plasma protein
( ) c. hemolysis

Renal damage can result from the release of *_____

_____.

— — — — — — — — — — — — — — —

a. X
b. —
c. X
free hemoglobin

110

Hemoconcentration is present in early burns, but after rehydration is established, the diminished number of erythrocytes becomes apparent.

Hemoconcentration is present (after/before) rehydration.

— — — — — — — — — — — — — — —

before

111

Fluid, which is frequently referred to as edema fluid, is collected at the burned area. Sodium enters the edema fluid in the burned area lowering the sodium content of the extracellular fluid. The low serum sodium may continue for weeks because of sodium loss to edema fluid and then later through diuresis.

Intracellular potassium is lost from the cells and is replaced by sodium. With oliguria (decreased urine output), serum potassium is elevated.

When oliguria occurs, the serum potassium is _____.
Serum sodium and cellular potassium are (increased/decreased) following a severe burn.

_ _ _ _ _ _ _ _ _ _ _ _ _ _ _ _

increased
decreased

112

Low serum sodium may continue for weeks due to sodium loss to

* _____ and through _____.

_ _ _ _ _ _ _ _ _ _ _ _ _ _ _ _

edema fluid
diuresis

113

Bicarbonate loss (HCO_3) accompanies loss of sodium.
Bicarbonate loss may be aggravated by starvation and retention of acid metabolic products by the kidneys.

As a result of a loss in bicarbonate, metabolic (acidosis/alkalosis) may exist.

_ _ _ _ _ _ _ _ _ _ _ _ _ _ _ _

acidosis

114
Metabolic acidosis can result from:
( ) a. loss of serum bicarbonate
( ) b. retention of acid metabolic products
( ) c. loss of erythrocytes

— — — — — — — — — — — — — — —

a. X, b. X, c. —

115
Though protein plasma levels are increased early in treatment due to the administration of plasma, the individual will continue to lose protein until healing occurs.

Low serum protein levels (will/will not) occur as tissue healing takes place.

— — — — — — — — — — — — — — —

will

116
A hazard in burn cases is infection. This will frequently delay the re-absorption of edema fluid from the site of the burn.

Surgical aseptic (sterile) techniques should be employed to reduce the

possibility of _____.

Infection will cause the edematous fluid to be reabsorbed (more slowly/more quickly).

— — — — — — — — — — — — — —

infection
more slowly

117

There is an increased capillary permeability to protein in the burned areas.

Uninjured areas (would/would not) have increased capillary permeability to protein.

— — — — — — — — — — — — — — —

would not

118

Indicate which of the following may occur as the result of burns:
( ) a. More water than protein will leave the blood vessels
( ) b. Low blood volume
( ) c. Increased number of erythrocytes
( ) d. Free-flowing hemoglobin
( ) e. Hemoconcentration before rehydration
( ) f. Decreased serum sodium
( ) g. Elevated serum potassium with hydration
( ) h. Decreased serum bicarbonate
( ) i. Metabolic alkalosis may occur
( ) j. Metabolic acidosis may occur
( ) k. Increased capillary permeability to protein in burned areas
( ) l. Increased osmotic pressure
( ) m. Elevated serum potassium with oliguria

— — — — — — — — — — — — — — —

a. X   h. X
b. X   i. —
c. —   j. X
d. X   k. X
e. X   l. X
f. X   m. X
g. —

Table 36 names the three degrees of burns, the affected tissues, and the characteristics of the burns. With first-degree burns, the epidermis or the outer layer of skin, is involved. With second-degree burns, the dermis or the true skin is involved and with third-degree burns, the subcutaneous tissues or fatty tissues are involved.

After studying this table carefully, proceed to the frames which follow. Refer to this table as needed.

TABLE 36   Degree of Burns and Their Characteristics

| Degree | Characteristics | | |
	Surface	Color	Pain
First degree (epidermis)	Dry, no blisters	Erythema (redness of skin)	Painful, hyperesthetic (very sensitive)
Second degree (dermis)	Moist, blisters	Mottled red	Painful, hypesthetic (less sensitive)
Third degree (subcutaneous tissues)	Dry	Pearly white or charred	Little pain; anesthetic (not sensitive)

Reference: *Therapeutic Notes,* October, 1963, page 236.

119

The three degrees of burns are:

1. _____

2. _____

3. _____

— — — — — — — — — — — — — — —

1. first degree
2. second degree
3. third degree

120

Define the following terms:

Epidermis._____

Dermis. _____

Subcutaneous tissue. _____
Give the names of the tissue involved according to the degree of the
burn.

First degree— _____

Third degree— _____

Second degree— _____

— — — — — — — — — — — — — — —

Epidermis. The outer layer of skin.
Dermis. True skin.
Subcutaneous tissue. Fatty tissue.
1st—epidermis
3rd—subcutaneous tissue
2nd—dermis

121

The characteristics commonly seen with first-degree burns are: the skin

surface is _____, the color is _____, and the

pain is _____ .

The characteristics commonly seen with second-degree burns are: the

skin surface is _____, the color is _____

_____, and the pain is _____

_____ .

The characteristics commonly seen with third-degree burns are: the skin

surface is _____, the color is _____

_____, and the pain is _____ .

– – – – – – – – – – – – – – –

dry
red (erythema)
hyperesthetic (very sensitive)
moist with blisters
mottled red
hypesthetic (less sensitive)
dry
pearly white or charred
anesthetic (not sensitive)

122

Place F.D. for first degree, S.D. for second degree, and T.D. for third degree burns as related to their characteristics:

_____ Skin surface is dry and pearly white or charred

_____ Skin surface is red and dry

_____ Skin surface is moist with blisters

_____ Painful and hypesthetic

_____ Little pain and anesthetic

_____ Painful and hyperesthetic

_____ Skin surface mottled red

_ _ _ _ _ _ _ _ _ _ _ _ _ _ _ _

T.D.   T.D.
F.D.   F.D.
S.D.   S.D.
S.D.

Diagram 12 explains the Rule of Nines in the estimation of amount and areas of body burns. The Rule of Nines uses 9% or multiples thereof in calculating the burned body surface. The five main regions in the estimation of burned surface are underlined. Be sure to know the five regions and the percentages of each. You may refer to this diagram as needed.

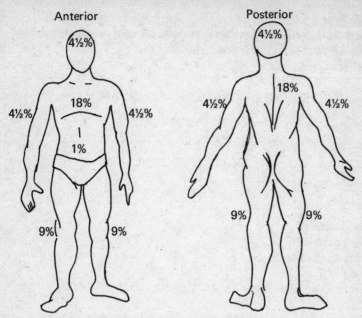

Anterior

4½%

18%

4½%   4½%

1%

9%   9%

Posterior

4½%

18%

4½%   4½%

9%   9%

Diagram 12. Rule of Nines for estimation of body surface.

Region	Percentage of Body Surface
Head and neck	9%
1. Anterior head and neck	
2. Posterior head and neck	
Upper extremities	18%
3. Right arm—anterior and posterior	
4. Left arm—anterior and posterior	
Trunk and buttocks	36%
5. Anterior surface	
6. Posterior surface	
Lower extremities	36%
7. Right leg and thigh—anterior and posterior	
8. Left leg and thigh—anterior and posterior	
9. Perineum and genitalia	1%

123

The Rule of Nines and the degree of burns are used in planning parenteral therapy for burned individuals.

What are the five main regions of the body used in the Rule of Nines?

1. _____

2. _____

3. _____

4. _____

5. _____

- - - - - - - - - - - - - -

Head and neck
Upper extremities
Trunk and buttocks
Lower extremities
Perineum and genitalia

124

In planning parenteral therapy for burns, the *_____

and *_____ are used.

The estimated percentage used for the anterior and posterior surfaces

of each arm is _____%.

The estimated percentage used for the anterior and posterior surfaces

of the head and neck is _____%.

- - - - - - - - - - - - - -

Rule of Nines
degree of burns
9%
9%

125

The estimated percentage used for the anterior surface of the trunk is

_____%.

The estimated percentage used for the anterior and posterior surfaces

of each thigh and leg is _____%.

— — — — — — — — — — — — — —

18%
18%

Review—Rule of Nines

Complete the percentages on the following chart:

Region	Percentage of Body Surface
Anterior head and neck	_____
Left anterior and posterior arm	_____
Anterior and posterior surfaces of the trunk and buttocks	_____
Anterior surface of the right thigh and leg	_____
Perineum and genitalia	_____
Anterior right arm	_____
Posterior trunk and buttocks	_____
Anterior and posterior surfaces of the lower extremities	_____

4–1/2%	1%
9%	4–1/2%
36%	18%
9%	36%

Table 37 differentiates between minor, moderate, and critical burns according to the degrees and percentage of burns. Know the three classifications of burns, the degrees, and percentages. Refer to the table as needed.

TABLE 37   Classification of Burns

Minor Burns	Moderate Burns	Critical Burns
First degree	Second degree of 15–30% of body surface	Second degree of over 30% of body
Second degree of less than 15% of body surface	Third degree of less than 10% of body surface except hands, face, feet, or genitalia	Third degree of more than 10% of body surface or of hands, face, feet, or genitalia
Third degree of less than 2% of body surface		

Reference: *Therapeutic Notes,* October, 1963, page 236.

126

The three classifications of burns are:

1. _____

2. _____

3. _____

— — — — — — — — — — — — — — —

1. minor burns
2. moderate burns
3. critical burns

127

To be classified as <u>minor burns</u>, the skin surface involved must be first

degree, or second degree of less than _____ % of body surface,

or third degree of less than _____ % of body surface.

_ _ _ _ _ _ _ _ _ _ _ _ _ _ _ _

15% body surface
 2% body surface

128

To be classified as <u>moderate burns</u>, the skin surface involved must be

second degree of _____ to _____ % of body surface, or

third degree of less than _____ % of body surface, provided that
the hands, face, feet, or genitalia are not burned.

_ _ _ _ _ _ _ _ _ _ _ _ _ _ _

15–30% of body surface
10% of body surface

129

To be classified as <u>critical burns</u>, the skin surface involved must be sec-

ond degree of over _____ % of body surface, or third degree of

more than _____ % of body surface, or of the hands, face,

_____ , or _____ .

_ _ _ _ _ _ _ _ _ _ _ _ _ _ _

30% of body surface
10% of body surface
feet or genitalia

130

Place the word <u>Minor</u> for minor burns, <u>Moderate</u> for moderate burns, and <u>Critical</u> for critical burns as they relate to the following statements:

_____ Burns of the hands, face, feet, or genitalia

_____ Third-degree burns of less than 10% of body surface

_____ Third-degree burns of more than 10% of body surface

_____ Second-degree burns of 15–30% of body surface

_____ Second-degree burns of over 30% of body surface

_____ Second-degree burns of less than 15% of body surface

_____ Third-degree burns of less than 2% of body surface

– – – – – – – – – – – – – – – –

critical
moderate
critical
moderate
critical
minor
minor

Review—Classification of Burns

Complete the following chart:

Minor Burns	Moderate Burns	Critical Burns
First degree		Second degree of over 30% of body surface
Second degree of less than 15% of body surface	Third degree of less than 10% of body surface except hands, face, feet, or genitalia	
		Hands, face, feet, or genitalia

_ _ _ _ _ _ _ _ _ _ _ _ _ _ _ _

Minor Burns
Third degree of less than 2% of body surface
Moderate Burns
Second degree of 15–30% of body surface
Critical Burns
Third degree of over 10% of body surface

131

Individuals with burns involving from 20–40% of body surface require careful and proper parenteral replacement therapy for survival.

In those with over 50% of the body surface involved, the mortality rate is high regardless of careful and proper parenteral therapy.

Individuals with 20–40% of body surface burns, having received careful and proper fluid replacement, have a (good/poor) prognosis, while those with 50% or more have a (good/poor) prognosis.

_ _ _ _ _ _ _ _ _ _ _ _ _ _ _ _

good
poor

Clinical Considerations

132

Various formulas have been devised and used as a basis for initiating therapy in the treatment of burns.

Cope and Moore state that accurate measured hourly urine flow is an important index for determination of adequate parenteral therapy. For all severely burned individuals, an in-dwelling catheter would be advisable to obtain *_____.

————————————————

hourly urine output

133

The desired rate of urine flow is 30–50 ml (cc) per hour.

Less than 25 ml of urine output per hour for an adult would indicate insufficient fluid intake or kidney dysfunction. Urine flow of over 100 ml per hour would indicate too much fluid intake.

For a burned individual, the desired urine output per hour is _____

_____.

Less than 25 ml per hour would indicate:
( ) a. too much fluids
( ) b. not enough fluids
( ) c. kidney dysfunction

————————————————

30–50 ml
a. —
b. X
c. X

134

If the low urine output is caused by insufficient fluids, then 5% dextrose in water or normal saline is used to increase the kidney function and urine output.

Failure to increase the amount of urine output by injection of hydrating solutions could indicate what? _____

_____.

Peritoneal dialysis or hemodialysis (artificial kidney) would have to be employed in severe cases where there is a(an) (increase/decrease) in urine flow over a period of time.

_ _ _ _ _ _ _ _ _ _ _ _ _ _ _ _

kidney damage or kidney dysfunction
decrease

135

Cope and Moore state in the management of severe burns, amount of

parenteral therapy should be determined by * _____

_____.

Urine flow less than 25 ml per hour is considered (normal/dangerous).

_ _ _ _ _ _ _ _ _ _ _ _ _ _ _ _

hourly urine output
dangerous

136

To determine whether poor urine output is due to renal damage or inadequate fluid intake, a Water Tolerance Test can be used. This test consists of giving 1000 ml of fluid in ½ hour.

A failure to note an increase in urine output would indicate * _____

_____.

_ _ _ _ _ _ _ _ _ _ _ _ _ _ _

renal damage

137

Individuals who are burned frequently complain of thirst. In order not to hydrate orally with plain water, the following oral liquid could be prepared in quenching thirst.

1 teaspoon of $NaHCO_3$ commonly known as soda
1 teaspoon of NaCl commonly known as salt
\} in one quart of cold water

One quart of this oral solution could be given per day on physician's orders

Massive amounts of plain water should not be given to prevent:

( ) a. edema
( ) b. water intoxication
( ) c. dehydration

The desired oral solution to be given for thirst consists of *_____

_____.

- - - - - - - - - - - - - - - -

a. −, b. X, c. −
one teaspoon of soda and one teaspoon of salt in one quart of water

138

Later, high protein liquids, between meals or at mealtime, are helpful for cell reconstruction. The following protein liquids could be used:

( ) a. eggnog
( ) b. ginger ale
( ) c. milkshake
( ) d. Coca Cola

- - - - - - - - - - - - - - - -

a. X, b. −, c. X, d. −

139

A very high hematocrit reading indicates _____ .

Many physicians prefer to maintain the hematocrit (Hct) at 45 or above the first 48 hours after the burn so that in rehydration, the hematocrit would:

( ) a. drop very low
( ) b. return to a normal range
( ) c. show a marked increase

— — — — — — — — — — — — — —

dehydration or hemoconcentration
a. —, b. X, c. —

140

The greatest fluid shift occurs during the first 8 hours after a burn and reaches its peak in 48 hours. Therefore, the critical period for fluid and electrolyte replacement is:

( ) a. the first 36 hours
( ) b. the first 48 hours
( ) c. the first 72 hours

— — — — — — — — — — — — —

a. —, b. X, c. —

141

After 48 hours, capillary permeability lessens, fluid reabsorption begins, and edema starts to subside. This is considered the stage of diuresis.

The stage of diuresis frequently begins after _____ .

During this stage, capillary permeability _____ ,

fluid _____ begins, and edema starts to _____

_____ .

— — — — — — — — — — — — —

48 hours   reabsorption
lessens   subside

142

After 48 hours, parenteral therapy is frequently restricted, providing the serum sodium and potassium levels are near normal.

Continuous parenteral therapy could result in overhydration. This could be hazardous since it could overload the circulation, causing pulmonary edema and cardiac failure.

After 48 hours, intravenous administration is _____.

Overloading the circulation can result in:

( ) a. pulmonary edema
( ) b. gastritis
( ) c. cardiac failure
( ) d. pancreatitis

_ _ _ _ _ _ _ _ _ _ _ _ _ _ _ _

decreased
a. X, b. −, c. X, d. −

143

The Brooke Army Hospital formula is frequently used in calculating the amount of parenteral therapy for the first 48 hours.

Parenteral therapy is not used for minor burns, but is used for moderate and _____ burns.

_ _ _ _ _ _ _ _ _ _ _ _ _ _

critical (severe)

The suggested fluid replacement for the first 48 hours after burns has been devised by the Brooke Army Hospital. Table 38 gives the Brooke Army Hospital formula for fluid replacement for the first and second 24 hours. A clinical example using this table is given in the frames that follow. Refer to this table as needed.

TABLE 38   Brooke Army Hospital Formula for Parenteral Therapy for the First 48 Hours after Burns

First 24 Hours	Second 24 Hours
0.5 ml (cc) colloid* per Kg of body weight × percent of total body burns	one-half the amount of colloid and electrolyte of the first 24 hours
1.5 ml (cc) electrolyte** per Kg of body weight × percent of total body burns	
2000 ml (cc) dextrose in water	2000 ml (cc) dextrose in water

*Colloid used: blood, dextran, plasma
**Electrolyte used: lactated Ringer's or normal saline.
1 Kg = 2.2 pounds
Note: Over 50% of body surface burns are calculated at 50% burns for fluid replacement purposes.

144
Solutions used for colloid replacement are _____,

_____ or _____.

Electrolyte solutions used are _____

_____ and _____.

– – – – – – – – – – – – – –

blood, dextran, or plasma
lactated Ringer's and normal saline

Clinical Example I

145

Mr. Greene weighed 154 pounds or 70 Kg had burned 30% of his body surface.

To estimate his fluid needs for the first 24 hours, one would calculate his fluid needs according to the Brooke's formula as:

0.5 cc colloid × 70 (Kg of body weight) × 30 (% of burned body sur-

face) = _____ cc of colloid to be given. [*Note:* When multiplying by 30, do not use the decimal point.]

1.5 cc electrolyte × 70 (Kg of body weight) × 30 (% of burned body

surface) = _____ of lactated Ringer's to be given.

Plus 2000 cc of dextrose in water

The total amount of parenteral fluids Mr. Greene should receive in the

first 24 hours following his burns would be _____.

– – – – – – – – – – – – – – – –

1050 cc
3150 cc
6200 cc

146

For the second 24 hours, according to Brooke's formula, Mr. Greene should receive:

_____ cc of colloid

_____ cc of lactated Ringer's

_____ cc of dextrose in water

The total amount of parental fluid for the second 24 hours would be

_____ cc.

– – – – – – – – – – – – – – –

525 cc
1575 cc
2000 cc
4100 cc

147

After 48 hours of parenteral therapy, Mr. Greene should receive a(an) (increase/decrease) in intravenous administration.

An increase of intravenous fluids after 48 hours could result in over-

hydration, which might lead to *_____

and _____ .

– – – – – – – – – – – – – – – –

decrease
pulmonary edema and cardiac failure

148

Over 50% of body surface burns are calculated as _____% burns for fluid replacement purposes.

– – – – – – – – – – – – – – – –

50%

149

During the first 24 hours, one-half of the fluids is given in the first 8 hours and the other half is given in the remaining 16 hours.

Mr. Greene should receive _____ cc of intravenous fluids the

first 8 hours and _____ cc the remaining 16 hours of the first day.

– – – – – – – – – – – – – – – –

3100 cc
3100 cc

Review—Brooke Army Hospital Formula

Complete the Brooke Army Hospital Formula for parenteral therapy
for the first 48 hours after a burn.

First 24 Hours

_____ cc colloid × Kg of

body weight × _____

_____ cc electrolyte ×

Kg of body weight × _____

2000 cc of dextrose/water

Second 24 Hours

_____ the amount of
colloid and electrolyte of the first
24 hours

_____ cc dextrose/
water

- - - - - - - - - - - - - - - -

First 24 Hours
0.5 cc; × % of body burns

1.5 cc; × % of body burns

Second 24 Hours
½ the amount

2000 cc

Clinical Example II

150

Mrs. Silver, age 35, received 25% of second- and third-degree body surface burns when her farm house caught on fire.

   The following areas of her body were burned:
   Face, 5%
   Right arm and hand, 9%
   Left arm, 5%
   Back and upper chest, 5%

Since the face, hand and upper chest were burned, Mrs. Silver would be
considered to have:
( ) a. Minor burns
( ) b. Moderate burns
( ) c. Critical burns

- - - - - - - - - - - - - - - -

a. −, b. −, c. X

151

Mrs. Silver's laboratory studies were:

Hemoglobin	13.5 grams
Hematocrit	44%
White Blood Count	20,300 cells per cu mm
Polymorphonuclear Cells (Polys)	65%

Venous Section (cutdown) was performed.

In the emergency room, she received:

Two injections of morphine sulfate: 1. gr. 1/6 (IM)
2. later gr. 1/6 (IV)

1000 cc normal saline with 2 million units of aq. penicillin I.V.
1000 cc normal saline with 5 million units of aq. penicillin I.V.
tetanus toxoid: 0.5 ml

Mrs. Silver received tetanus toxoid since she was subject to infection by anaerobic microbes, such as *Clostridium tetani.*

She received morphine sulfate for _____.
She received aq. penicillin since her:
( ) a. hemoglobin was elevated
( ) b. hematocrit was elevated
( ) c. W.B.C. was elevated

– – – – – – – – – – – – – – –

relief of pain
a. –
b. –
c. X

The laboratory studies of Mrs. Silver, given in Table 39, show how her results deviate from the norms at the time of her illness.

TABLE 39 Laboratory Studies for Mrs. Silver

Laboratory Tests*	On Admission	1st Day	2nd Day	3rd Day	4th Day	5th Day	6th Day	7th Day
Hematology								
Hemoglobin (12.9–17.0 grams)	13.5	17.6						
Hematocrit (40–46%)	44	49 54	56 52 61 64	55	51	46	36	35
W.B.C. (white blood count) (5,000–10,000 cu mm)	23,000		11,658					
Biochemistry								
B.U.N. (blood urea nitrogen) (10–15 mg/100 ml)	11	14						
Plasma CO₂ (50–70 vol %) / (22–32 mEq/L)	$\frac{30}{14}$	$\frac{44}{20}$	$\frac{44}{20}$		$\frac{57}{26}$		$\frac{57}{26}$	
Plasma chloride (98–107 mEq/L)	105	105	105		97		103	
Plasma sodium (135–146 mEq/L)	141	137	137		134		141	
Plasma potassium (3.5–5.3 mEq/L)	4.3	4.8	4.8		4.2		4.4	

*Laboratory norms from Wilmington Medical Center, Delaware Division.

152

Mrs. Silver's plasma chloride, sodium, and potassium (were/were not) in normal range.

Due to her low plasma CO_2 on admission, she would be in a state of

metabolic _____.

– – – – – – – – – – – – – – – –

were
acidosis

153

In the above frame you noted that Mrs. Silver's plasma electrolytes were in normal range. Can you recall the normal range for the following electrolytes without referring to Table 39?

Plasma chloride. _____

Plasma sodium. _____

Plasma potassium. _____

– – – – – – – – – – – – – – – –

Cl– 98–107 mEq/L
Na– 135–146 mEq/L
K– 3.5–5.3 mEq/L

154

Her elevated hematocrit during the first 48 hours following admission

would be an indication of _____.

– – – – – – – – – – – – – – –

dehydration. You may have answered hemoconcentration, O.K.

155

Her elevated W.B.C. indicated:

( ) a. an increased number of   ( ) b. infection
         white blood cells   ( ) c. head cold

– – – – – – – – – – – – – – – –

a. X, b. X, c. –

Clinical Management

156

Mrs. Silver weighed 65 Kg and had a total of 25% body burns. Calculate the following for fluid replacement:

0.5 × 65 × 25 = _____ cc colloid

1.5 × 65 × 25 = _____ cc electrolyte

_____2000_____ cc dextrose/water

Total   _____ cc for first 24 hours

During the first 8 hours Mrs. Silver received one-half of the intravenous

fluid. This amount would be _____ cc.
She received:
    1000 cc normal saline
     350 cc plasma
    1000 cc normal saline
     250 cc blood
    2600 cc for the first 8 hours
Note: Mrs. Silver did not receive lactated Ringer's because her electrolytes were not low. You will notice in this program and in the clinical area that the figures are rounded off.
    Then she received:
        500 cc saline
        250 cc plasma
        400 cc 5% dextrose in water
     1150 cc for the second 8 hours
    Later she received:
        1500 cc 5% dextrose in water for the third 8 hours

The total amount of colloid she received was _____.

The total amount of electrolyte she received was _____.

The total amount of dextrose in water she received was _____.

- - - - - - - - - - - - - - - - -

812.5 cc	850 cc
2437.5 cc	2500 cc
5250 cc	1900 cc
2600 cc	

157

Check the following amounts of fluids she should receive during the
second 24 hours according to Brooke's formula.

( ) a.   800 cc colloid
( ) b. 1250 cc electrolyte
( ) c. 1000 cc 5% dextrose in water
( ) d.   425 cc colloid
( ) e. 1500 cc electrolyte
( ) f. 2000 cc 5% dextrose in water
( ) g. 2600 cc total for second 24 hours
( ) h. 3675 cc total for second 24 hours
( ) i. 4225 cc total for second 24 hours

– – – – – – – – – – – – – – – –

a. –   f. X
b. X   g. –
c. –   h. X
d. X   i. –
e. –

158

In Mrs. Silver's case, urine output was measured hourly and tested for
specific gravity.

After the first 8 hours, Mrs. Silver's urine output was 250 cc per hour,
so the I.V. fluid flow rate was cut back.

During the third 8 hours, her urine output had fallen to 5-10-15 cc per
hour. The fluid flow rate should be (increased/decreased).

In a case like Mrs. Silver's, which of these intravenous fluids would be
the preferred ones to give when there is a decrease in urine output.

( ) a. blood
( ) b. normal saline
( ) c. 5% dextrose in water
( ) d. 10% dextrose in saline

– – – – – – – – – – – – – – –

increased
a. –   c. X
b. X   d. –

159

The specific gravity of Mrs. Silver's urine ranged from 1.005–1.017. Specific gravity of urine is the weight (waste products) in relationship to water, 1.000. Specific gravity norm for urine is (1.010–1.030).

When Mrs. Silver's specific gravity was 1.005 and 1.008, there would be (more/less) waste products in her urine than the norm.

The waste products would be more concentrated in her _____.

— — — — — — — — — — — — — — —

less
plasma

Review—Burns

Q. 1

After a burn, there is a shift of fluid and electrolytes from the plasma to

the *_____.

— — — — — — — — — — — — — — —

burned area or the interstitial space

Q. 2

With burns, more water than protein will leave the plasma, thus (increasing/decreasing) the plasma osmotic pressure.

— — — — — — — — — — — — — — —

increasing

Q. 3

Erythrocytes are destroyed due to _____ and

_____ into the burned area.

— — — — — — — — — — — — — — —

hemolysis and bleeding

Q. 4

With oliguria, as a result of burns, there is a(an) (increase/decrease) in serum potassium.

_ _ _ _ _ _ _ _ _ _ _ _ _ _ _

increase

Q. 5

Sodium deficit is present since sodium leaves the plasma and enters the

* _____.

_ _ _ _ _ _ _ _ _ _ _ _ _ _ _

edema fluid in the burned area

Q. 6

The serum bicarbonate is lost, causing a decrease in plasma _____

and resulting in a state of metabolic _____.

_ _ _ _ _ _ _ _ _ _ _ _ _ _

CO_2
acidosis

Q. 7

There is an increased capillary permeability to protein only in the

_____ areas.

_ _ _ _ _ _ _ _ _ _ _ _ _ _

burned

Q. 8

In planning parenteral therapy for burned individuals, the *_____

_____ and the *_____

_____ are used.

_ _ _ _ _ _ _ _ _ _ _ _ _ _ _ _

Rule of Nines and degree of burns (also urine output and Brooke's formula)

Q. 9

Less than 25 ml of urine output per hour for an adult would indicate

*_____

or *_____.

_ _ _ _ _ _ _ _ _ _ _ _ _ _ _ _

insufficient fluid intake or kidney dysfunction

Q. 10

Cope and Moore state that an important index for the determination of

adequate parenteral therapy is an accurate record of *_____

_____.

measured hourly urine output

Q. 11

An oral liquid which can be given in most burned situations, as ordered

by the physician, for thirst is *_____

_____

_____.

_ _ _ _ _ _ _ _ _ _ _ _ _ _ _

one teaspoon of salt and one teaspoon of soda in one quart of cold
water

Q. 12
In burns, the greatest fluid shift occurs within _____ .

_ _ _ _ _ _ _ _ _ _ _ _ _ _ _

48 hours

Q. 13
The stage of diuresis occurs after 48 hours following a burn when capil-

lary permeability _____, fluid *_____

_____, and edema *_____

_____.

_ _ _ _ _ _ _ _ _ _ _ _ _ _ _

lessens (decreases)
reabsorption begins
starts to subside

Q. 14
With burns, overhydration is hazardous since it could overload the cir-

culation, causing *_____ and

*_____.

_ _ _ _ _ _ _ _ _ _ _ _ _ _

pulmonary edema and cardiac failure

Q. 15
Solutions used for colloid replacement are _____, _____,

or _____.

_ _ _ _ _ _ _ _ _ _ _ _ _ _ _

blood, dextran, or plasma

Q. 16

Electrolyte solutions used are _____

and _____ .

_ _ _ _ _ _ _ _ _ _ _ _ _ _ _

lactated Ringer's and normal saline

Q. 17

Mrs. Silver's elevated hematocrit reading the first 48 hours was an indication of _____ .

_ _ _ _ _ _ _ _ _ _ _ _ _ _ _

dehydration

DIABETIC ACIDOSIS

Physiological Factors

160

The inability of the body to utilize glucose results in an increasing concentration of sugar in the blood.

The elevation of blood sugar increases glucose concentration in the glomerular filtrate of the kidneys. This increased solute load in the kidneys requires extra fluid from the rest of the body; therefore, polyuria (increased amount of urine) occurs.

In diabetic acidosis, explain in your words why polyuria occurs.

_____

_____

_____

_____

_ _ _ _ _ _ _ _ _ _ _ _ _ _ _

An increased glucose concentration will cause an increase in intravascular fluid so that excess glucose can be excreted, or a similar response.

161

When the concentration of glucose in the glomerular filtrate exceeds the renal threshold for tubular reabsorption, glycosuria (sugar in the urine) results.

This increased glucose concentration in the urine results in osmotic diuresis, because the increased osmotic pressure in the urine draws

_____ from the body.

_ _ _ _ _ _ _ _ _ _ _ _ _ _ _ _

fluid

162

Osmotic diuresis inhibits reabsorption of sodium causing excessive sodium to be washed out with glycosuria.

An increased blood sugar will cause:
( ) a. osmotic diuresis
( ) b. polyuria
( ) c. glycosuria
( ) d. sodium excretion
( ) e. sodium reabsorption
( ) f. fluid reabsorption

_ _ _ _ _ _ _ _ _ _ _ _ _ _ _ _

a. X
b. X
c. X
d. X
e. —
f. —

163

Elevated blood sugar will also increase the hypertonicity of the extra-cellular fluid.

The hypertonicity of the extracellular fluid leads to a withdrawal of fluid from the cells, thus equalizing the extracellular and intracellular osmolality.

The fluid from the cells will dilute the extracellular sodium concentration, producing (hypernatremia/hyponatremia).

The migration of intracellular fluid into the extracellular fluid will result in:

( ) a. cellular dehydration
( ) b. cellular hydration

_ _ _ _ _ _ _ _ _ _ _ _ _ _ _ _

hyponatremia
a. X
b. —

164

Failure to metabolize glucose leads to an increased fat utilization for energy producing ketonic acids (products of fat catabolism) which accumulate in the blood. Ketosis then occurs, which is an excess number of ketone bodies in the blood.

Failure of glucose metabolism will cause:

( ) a. fat utilization for energy
( ) b. excessive ketone bodies due to fat catabolism

Excessive number of ketone bodies in the body is known as _____.

Ketosis will lead to metabolic _____.

_ _ _ _ _ _ _ _ _ _ _ _ _ _ _

a. X
b. X
ketosis
acidosis

165

In renal excretion, the ketones (strong acids) will combine with the cation, sodium, causing a sodium depletion. Acetone bodies are simple
ketones that are excreted as acetonuria.

The additional solute load of ketones in the glomerular filtrate will result in:
(   ) a. a decreased loss of water in the formation of acetonuria
(   ) b. a greater loss of water in the formation of acetonuria

— — — — — — — — — — — — — — —

a. —
b. X

166

Polyuria can result from:
(   ) a. glycosuria
(   ) b. anuria
(   ) c. acetonuria

With the loss of water, the concentration of the blood _____

and the blood volume _____.

— — — — — — — — — — — — — — —

a. X
b. —
c. X
increases
decreases

167

As a result of metabolic acidosis, nausea and vomiting occur and will decrease the fluid and electrolytes.

There is an increase of water loss by way of the lungs due to Kussmaul breathing (hyperventilating—deep, rapid breathing).

Dehydration occurs from:
( ) a. nausea and vomiting
( ) b. Kussmaul breathing
( ) c. oliguria
( ) d. polyuria

_ _ _ _ _ _ _ _ _ _ _ _ _ _ _

a. X
b. X
c. —
d. X

168

The failure of cellular utilization of glucose permits potassium to leave the cells. The serum potassium may therefore show normal or high volume.

There is a serum potassium loss due to vomiting and renal excretion, but the hemoconcentration can cause the serum potassium to show

*_____.

_ _ _ _ _ _ _ _ _ _ _ _ _ _

normal or high volume

169

With the increasing dehydration and hemoconcentration, the renal blood flow will become inadequate and the excretion of potassium will be lessened. Therefore, the blood levels of potassium will be _____

_____.

_ _ _ _ _ _ _ _ _ _ _ _ _ _

elevated

Table 40 shows the distribution of electrolytes in plasma or serum normally, and in early and late stages of diabetic acidosis. Note in particular the increase in ketone bodies (products of fat catabolism) in late acidosis. Know what happens to the bicarbonate ions when there is an increase in ketones. You are not expected to know the effects of the cations. Refer to this table as needed.

TABLE 40   Serum Electrolytes in Diabetic Acidosis

Normal	Acidosis (Early stage)	Acidosis (Late Stage)
Na^+	HCO_3	HCO_3
HCO_3-	Ketones	Ketones
Cl^-	Cations: Na^+, K^+, Ca^{++} and Mg^{++}	Cl^-	Cations: Na^+, K, Ca^{++} and Mg^{++}	Cl^-
PO_4, SO_4^{--}	Urinary Excretory Products	Urinary Excretory Products
K^+		
Ca^{++}	Protein	
Mg^{++}		

Reference: Harry Statland, *Fluid and Electrolytes in Practice*, J.B. Lippincott Co., Philadelphia, Pa., p. 268.

170

The center column of Table 40 shows that the ketones increase at the expense of the *_____.

Because of dehydration, the serum osmotic pressure is _____

and the serum levels of electrolytes will be _____.

— — — — — — — — — — — — — — — —

bicarbonate ions
increased
increased

171

With an increased number of ketones (Table 40, last column), the alkali reserve (bicarbonate) will be _____.

— — — — — — — — — — — — — — — —

decreased

172

With a decrease in the number of bicarbonates and an increase in ketones which are acid, what will occur—metabolic alkalosis or metabolic acidosis? _____.

— — — — — — — — — — — — — — — —

metabolic acidosis

Clinical Considerations

173

Dehydration is one of the major symptoms and concerns for individuals in diabetic acidosis.

When there is a marked intracellular and extracellular fluid depletion, the end result is:

( ) a. decreased hemoconcentration
( ) b. increased hemoconcentration
( ) c. decreased blood volume
( ) d. increased blood volume

_ _ _ _ _ _ _ _ _ _ _ _ _ _ _

a. –
b. X
c. X
d. –

Table 41 gives the classic signs and symptoms of dehydration of patients in diabetic acidosis. Actually you may not find a patient displaying all the listed laboratory results or all the listed observed symptoms when dehydration is present, but this table should serve as a guide in recognizing dehydration in diabetic patients.

There is no horizontal relationship between the two columns in the table. Take time in studying this table. Quiz yourself on the signs and symptoms and then proceed to the frames. Refer to the table as needed.

TABLE 41 Signs and Symptoms of Dehydration in Diabetic Acidosis

Laboratory Test Results	Observed Symptoms
Hyperglycemia	Thirst
Glycosuria (glucose diuresis)	Polyuria due to
	Glycosuria
Acetonuria (ketone diuresis)	Acetonuria
Elevated hematocrit	Concentrated urine
Elevated leukocyte count	Nausea and vomiting
Decrease in CO_2 combining power or bicarbonate resulting in metabolic acidosis	Sweating
	Blood pressure low, and a rapid, thready pulse due to hypovolemia
Bicarbonate ions are replaced with ketones which are strong acids	Kussmaul breathing (hyperventilation)— respiratory mechanism for lowering serum acid
Serum pH 6.9—acidosis	
Hypochloremia—excreted with sodium and water	Lack of intake due to nausea and vomiting
Hypomagnesemia and hypophosphoremia due to cellular dehydration	Abdominal cramps
	Dry mucous membranes
Hyponatremia	
	Skin dry and lips dry and parched
	Sunken eyes and soft eyeballs

174

Check the laboratory findings of dehydration in diabetic acidosis:

( ) a. hypoglycemia
( ) b. hyperglycemia
( ) c. hyponatremia
( ) d. hyperchloremia
( ) e. decreased CO_2 combining power
( ) f. hypovolemia
( ) g. low hematocrit count
( ) h. elevated leukocyte count
( ) i. glycosuria
( ) j. acetonuria
( ) k. serum pH 6.8
( ) l. hypermagnesemia
( ) m. elevated hematocrit count

– – – – – – – – – – – – – – –

a. –
b. X
c. X
d. –
e. X
f. X
g. –
h. X
i. X
j. X
k. X
l. –
m. X

175

Check the physical symptoms of dehydration in diabetic acidosis:

( ) a. abdominal cramps
( ) b. thirst
( ) c. abdominal distention
( ) d. polyuria
( ) e. nausea and vomiting
( ) f. Kussmaul breathing
( ) g. dyspnea
( ) h. increased blood pressure
( ) i. pulse fast and thready
( ) j. dry mucous membrane
( ) k. moist skin
( ) l. sunken eyes and soft eyeballs

_ _ _ _ _ _ _ _ _ _ _ _ _ _ _ _ _

a. X
b. X
c. —
d. X
e. X
f. X
g. —
h. —
i. X
j. X
k. —
l. X

176

Check the following laboratory findings and symptoms that relate to dehydration in diabetic acidosis:

( ) a. hypernatremia
( ) b. hyperglycemia
( ) c. thirst
( ) d. nausea and vomiting
( ) e. abdominal cramps
( ) f. bradycardia—slow pulse
( ) g. polyuria
( ) h. decreased blood pressure
( ) i. hypochloremia
( ) j. hypomagnesemia
( ) k. hyperphosphoremia
( ) l. decreased CO_2 combining power
( ) m. hypovolemia
( ) n. elevated hematocrit
( ) o. elevated leukocyte count
( ) p. sunken eyes and soft eyeballs
( ) q. decreased leukocyte count
( ) r. Kussmaul breathing

— — — — — — — — — — — — — — —

a. —	j. X
b. X	k. —
c. X	l. X
d. X	m. X
e. X	n. X
f. —	o. X
g. X	p. X
h. X	q. —
i. X	r. X

177

In the first 24 hours, 80% of the total water and salt deficit should be replaced.

For the other electrolytes, there is less urgency since the rate of assimilation of the intracellular electrolytes is limited. Administration of potassium must be included, but not in early treatment (unless indicated) since serum potassium may be increased to toxic level.

With diabetic acidosis, 80% of (salt and water/potassium and magnesium) deficit should be replaced in the first 24 hours.

Cellular assimilation of electrolytes is (faster/slower) than extracellular.

— — — — — — — — — — — — — — — —

salt and water
slower

Table 42 outlines the restoration for extracellular fluid deficit and intracellular fluid deficit. Actually ECF is restored directly from I.V. therapy whereas ICF follows from I.V. therapy on the basis of internal changes. Study the table carefully, noting when the listed elements are restored. Refer to the table as needed.

TABLE 42   Restoration of ECF and ICF Deficits

Restoration of Extracellular Fluid Deficits	Restoration of Intracellular Fluid Deficits
Large deficiencies may be made up in 24 to 48 hours. Na Cl HCO_3 Water	As the serum osmolality drops with glucose utilization, fluid will enter the cells. This frequently occurs in the first two days. From around the 6th to the 10th days, K, Mg, PO_4, and water are replenished in the cells. Around the 7th day, nitrogen replacement begins.

178

Restoration of intracellular fluid deficit is somewhat _____ than extracellular fluid deficit.

_ _ _ _ _ _ _ _ _ _ _ _ _ _ _

slower

179

Around the 6th to 10th days, the electrolytes _____,

_____ , and _____ are replenished in the cells.

Nitrogen replacement begins around the_____ day.

_ _ _ _ _ _ _ _ _ _ _ _ _ _ _

potassium, magnesium, and phosphate
7th

Review—Restoration of ECF and ICF Deficits

Complete the times on the chart:

Restoration ECF	Restoration ICF
Large deficiencies, Na, Cl, HCO_3 and water may be made up in _____ hours.	Fluids will reenter the cells in the _____.
	K, Mg, PO_4, and water replenish the cells _____ _____.
	Nitrogen replacement begins the _____ _____.

- - - - - - - - - - - - - - - -

Restoration ECF
24 to 48 hours

Restoration ICF
1. first 2 days
2. around the 6th to the 10th day
3. 7th day

180

The longer the acidosis persists, the more resistant the individual is likely to be to insulin. Insulin is a hormone secreted by the pancreas and is essential for oxidation and utilization of blood sugar (glucose).

Only regular, unmodified, or crystalline insulin may be administered in intravenous fluids.

If acidosis persists, the individual may require (more/less) insulin.

The following types of insulin can be administered intravenously:
( ) a. NPH
( ) b. regular
( ) c. unmodified
( ) d. PZI
( ) e. crystalline

- - - - - - - - - - - - - - - -

more
a. –, b. X, c. X, d. –, e. X

181

Individuals who are ill with diabetes are advised to go to bed since rest reduces metabolism, and fat and protein catabolism would be less likely to occur.

These individuals should also be protected from overheating and chilling. If they are in a state of vascular collapse, extra heat should not be applied since it will increase vasodilatation and intensify the failure of the circulation.

Rest reduces metabolism in an ill diabetic individual; therefore, rest will decrease the chance of _____ and _____ catabolism.

In the state of vascular collapse, extra heat may cause further _____ _____.

_ _ _ _ _ _ _ _ _ _ _ _ _ _ _

fat and protein
vasodilatation

182

Orange juice and broth are the fluids which are frequently given after nausea and vomiting have ceased.

Broth is a good source of sodium, potassium, and water, and is usually tolerated when served hot.

Orange juice is also a good source of potassium as well as sugar and water.

Ginger ale is also helpful after cessation of nausea and vomiting.

The oral fluids which are frequently tolerated soon after nausea and vomiting are:
( ) a. broth
( ) b. milkshake
( ) c. orange juice
( ) d. ginger ale

Broth is rich in the following fluid and electrolytes:
( ) a. calcium
( ) b. potassium
( ) c. water
( ) d. sodium
( ) e. sulfate

Orange juice is rich in:
( ) a. sugar
( ) b. potassium
( ) c. calcium
( ) d. water

— — — — — — — — — — — — —

a. X	a. −	a. X
b. −	b. X	b. X
c. X	c. X	c. −
d. X	d. X	d. X
	e. −	

183

Individuals being treated for diabetic acidosis should be observed closely for hypoglycemic reaction. This is especially true when frequent large doses of regular, unmodified, or crystalline insulin are given. Hypoglycemic reactions resemble shock symptoms.

Large doses of insulin could cause a hypoglycemic reaction since insulin (increases/lowers) the blood sugar.

_ _ _ _ _ _ _ _ _ _ _ _ _ _ _ _

lowers

Clinical Example

184

Mrs. Thompson arrived in the emergency room in a semi-comatose state of consciousness. Prior to admission, she had been vomiting and had complained of being weak. The family stated she had had a severe cold with a fever for weeks. They felt the vomiting was due to a viral infection.

The mucosa in her mouth was dry. Vomiting and dry mucosa would

indicate _____ .

The respirations were rapid and deep, which would be an indication of:
( ) a. Kussmaul breathing
( ) b. dyspnea

Her heart sinus rhythm was sinus tachycardia (pulse rate 120). Her breath had a very sweet smell.

The family stated she did not have diabetes mellitus, but there was a history of it in their family.

In the emergency room, a stat blood chemistry was done and a retention catheter was inserted.

The blood sugar was 476 mg/100 ml, the norm being 80–120 mg/100 ml. This would indicate a (hypoglycemic/hyperglycemic) state.

The plasma CO_2 combining power was very low, which would indicate

an _____ state.

– – – – – – – – – – – – – – – –

dehydration
a. X
b. –
hyperglycemic
acidotic

Table 43 gives the laboratory studies of Mrs. Thompson which show how her results deviate from the norms at the time of her illness.

TABLE 43 Laboratory Studies for Mrs. Thompson

Laboratory Tests*	On Admission		1st Day		2nd Day	3rd Day
Hematology						
Hemoglobin (12.9–17.0 grams)	17.8					
Hematocrit (40–46%)	52					
Biochemistry						
B.U.N. (blood urea nitrogen) (10–15 mg/100 ml)	15					
Sugar feasting (under 150 mg/100 ml)	476	825	458	382	70	144
Acetone	$\dfrac{+1}{1:10}$	$\dfrac{+1}{1:10}$	$\dfrac{\text{Trace}}{1:8}$			
Plasma CO_2 $\dfrac{(50–70 \text{ vol \%})}{(22–32 \text{ mEq/L})}$	$\dfrac{7}{3}$	$\dfrac{10}{4}$	$\dfrac{14}{6}$	$\dfrac{18}{8}$	$\dfrac{34}{15}$	$\dfrac{44}{20}$
Plasma chloride (98–107 mEq/L)	104		130	132	133	110
Plasma sodium (135–146 mEq/L)	137		151	159	164	145
Plasma potassium (3.5–5.3 mEq/L)	4.8		2.7	3.2	4.2	4.5

*Laboratory norms from Wilmington Medical Center, Delaware Division.

185

Her urinalysis was as follows:

color — dark yellow
specific gravity — 1.024
reaction — acid
albumin — +3
sugar — +4
W.B.C. — many

Her specific gravity shows (refer to frame 159 if needed):
( ) a. a very high range
( ) b. a high average range
( ) c. a low range
( ) d. an indication of an increased amount of products in the urine

The +4 sugar in the urine would indicate (hypoglycemia/hyperglycemia).

The +3 albumin in the urine would indicate:
( ) a. normal range
( ) b. pathological involvement, i.e., kidney cell injury

— — — — — — — — — — — — — — — —

a. —
b. X
c. —
d. X
hyperglycemia
a. —
b. X

186

Mrs. Thompson's hemoglobin and hematocrit counts were:

( ) a. normal
( ) b. below normal
( ) c. above normal
( ) d. an indication of mild edema
( ) e. an indication of mild dehydration

_ _ _ _ _ _ _ _ _ _ _ _ _ _ _

a. —
b. —
c. X
d. -
e. X

187

The feasting blood sugars (blood drawn after eating) on admission and the first day were:

( ) a. normal
( ) b. below normal
( ) c. above normal
( ) d. an indication of hyperglycemia
( ) e. an indication of hypoglycemia

The second day, her blood sugar was 70 mg/100 ml, which would indi-

cate a _____reaction.

_ _ _ _ _ _ _ _ _ _ _ _ _ _ _

a. —
b. —
c. X
d. X
e. —
hypoglycemic

188

Mrs. Thompson's plasma CO_2 combining power was (normal/very low/ very high).

Her CO_2 combining power would indicate:
( ) a. metabolic acidosis
( ) b. metabolic alkalosis
( ) c. a bicarbonate loss
( ) d. a bicarbonate increase

— — — — — — — — — — — — — — —

very low
a. X
b. —
c. X
d. —

189

Mrs. Thompson's plasma chloride, sodium and potassium on admission were in (high/low/normal) range.

On the first day, the laboratory studies indicated:
( ) a. hyperchloremia
( ) b. hypochloremia
( ) c. hypernatremia
( ) d. hyponatremia
( ) e. hyperkalemia
( ) f. hypokalemia

— — — — — — — — — — — — — — —

normal range
a. X
b. —
c. X
d. —
e. —
f. X

190

On admission, Mrs. Thompson's laboratory results were hemoglobin 17.8, hematocrit 52%, B.U.N. 15 mg/100 ml, feasting blood sugar 476, acetone +1 in 1:10 dilution, plasma CO_2 7% and 3 mEq/L, plasma chloride 104 mEq/L, plasma sodium 137 mEq/L, and plasma potassium 4.8 mEq/L.

Indicate whether these results are high, low, or normal. If high or low, give one condition each abnormality might indicate.

hematocrit _____

hemoglobin _____

B.U.N. _____

feasting blood sugar _____

acetone _____

plasma CO_2 _____

plasma chloride _____

plasma sodium _____

plasma potassium _____

– – – – – – – – – – – – – – –

high, dehydration	high, hyperglycemia	normal range
high, dehydration	high, fat catabolism	normal range
normal range	low, metabolic acidosis	normal range

191

What state is indicated by each of the following results:
The first day her plasma chloride was 130–132 mEq/L, her plasma sodium was 151–159 mEq/L, plasma potassium was 2.7–3.2 mEq/L.

plasma chloride _____

plasma sodium _____

plasma potassium _____

– – – – – – – – – – – – – – –

hyperchloremia   hypernatremia   hypokalemia

Clinical Management

Table 44 explains the medical treatment for the first 3 days for Mrs. Thompson which includes parenteral therapy, electrolytes and insulin. The amount of electrolytes she received varied according to her serum electrolytes. Also her insulin varied according to her blood sugar. On the second day she did not receive insulin, because blood sugar was low.

   Study this table carefully and then proceed to the frames. Refer to the table as needed.

192

In diabetic acidosis, there is frequently a serum sodium decrease before treatment due to:
( ) a. fluid intake
( ) b. vomiting
( ) c. urine excretion

In the emergency room, Mrs. Thompson received saline, $NaHCO_3$, and insulin.

The sodium bicarbonate was to:
( ) a. increase the plasma CO_2 combining power
( ) b. decrease the plasma CO_2 combining power
( ) c. increase the bicarbonate in the plasma
( ) d. decrease the bicarbonate in the plasma

— — — — — — — — — — — — — — — —

a. —
b. X
c. X
a. X
b. —
c. X
d. —

TABLE 44 Fluid, Electrolytes, and Insulin Replacements for Mrs. Thompson

Date	Parenteral Therapy	Electrolytes	Insulin
Emergency room	2000 cc normal saline	Na HCO₃ ampules (sodium bicarbonate)	Regular insulin U.50 subcutaneously
1st Day	1000 cc 5% dextrose/saline	KCl 40 mEq (potassium chloride) NaHCO₃ amp 1	1. regular insulin U.50 subcutaneously 2. regular insulin U.50 I.V. 3. One hour later:
	1000 cc normal saline	NaHCO₃ amp 1 KCl 60 mEq	regular insulin U.50 subcutaneously qh × 4 (total 200 U. in 4 hours) 4. Four hours later:
2nd Day	1000 cc 5% dextrose in water	NaHCO₃ amp 1 KCl 80 mEq	regular insulin U.50 subcutaneously regular insulin U.25 I.V.
	In 4 hours 500 cc 2.5% dextrose in water	NaHCO₃ amp 1	
	In 4 hours 500 cc 2.5% dextrose in water		
	Keep vein open with 500 cc 2.5% dextrose in water		
3rd Day	Oral fluids		N.P.H. insulin U.20 to start

193

The first day she was given potassium chloride because *_____

_____ .

She was given regular insulin U.50 on an hourly basis. This was to:

( ) a. lower her elevated blood sugar
( ) b. increase her blood sugar
( ) c. utilize the free glucose in her blood circulation

_ _ _ _ _ _ _ _ _ _ _ _ _ _ _

her plasma potassium was low
a. X
b. —
c. X

194

The second day she did not receive normal saline I.V. This is because

*_____ and

_____ .

She did not receive insulin the second day since her blood sugar was

_____ .

_ _ _ _ _ _ _ _ _ _ _ _ _ _ _

her plasma sodium and plasma chloride were elevated
low

195

Mrs. Thompson received normal saline to replace the _____

and _____ losses due to _____ and

_____.

She received $NaHCO_3$ since her plasma CO_2 was _____.

She received KCl since her plasma potassium was _____.

She received regular insulin to _____

and _____.

She also received additional sodium and chloride from the electrolyte

drugs _____ and _____.

– – – – – – – – – – – – – –

sodium and chloride
vomiting and urine excretion
(very)low
(very)low
lower her blood sugar and to metabolize the glucose
$NaHCO_3$ and KCl

Table 45 outlines selected nursing interventions and rationale in caring
for Mrs. Thompson or for any individual receiving replacement therapy
as a result of diabetic acidosis. Actually this table can be used as a reference table in reviewing the major symptoms of electrolyte and glucose
imbalance. (Diabetic acidosis is the same as metabolic acidosis, except
that the cause is diabetes.) You may wish to refer to this table frequently.

TABLE 45 Selected Nursing Interventions and Rationale in Caring for the Diabetic Acidotic Patient during Replacement Therapy

Selected Nursing Interventions	Rationale
Prepare the intravenous fluids, sodium bicarbonate, potassium chloride, and insulin according to physician's orders.	This type of medical treatment is frequently used in the care of an individual in a diabetic acidotic state for replacement of fluid and electrolyte losses.
Know the action, dosage, and toxic effects of the drugs and solutions being administered.	This knowledge is absolutely important before their administration in order to ward off any errors from misinterpretations and for observing side effects and adverse reactions.
Observe the patient for adverse reactions from parenteral and drug therapy, i.e., hypokalemia, hyperkalemia, hyponatremia, hypernatremia, metabolic acidosis, metabolic alkalosis, hypoglycemia, hyperglycemia, hypovolemia, hypervolemia, and renal functions.	Observation by the nurse will aid in detecting the adverse reactions from the drug therapy.

Hypokalemia	Hyperkalemia
Malaise	Abdominal cramps
Abdominal distention	Tachycardia then bradycardia
Arrhythmia	Weakness, numbness or tingling in extremities
Hypotension causing dizziness	5.3 mEq/L and increase
Muscular weakness	
Mental depression	
Drowsiness	
Confusion	
Paralysis of respiration	
3.5 mEq/L and decrease	

Hyponatremia	Hypernatremia
Abdominal cramps	Flushed skin
Muscular weakness	Rough, dry tongue
Headaches	Vital signs:
Nausea and vomiting	B.P. low
135 mEq/L and decrease	Pulse rapid
	Temperature high
	146 mEq/L and increase

Selected Nursing Interventions		Rationale
Metabolic Acidosis	Metabolic Alkalosis	
*Plasma CO_2 low	Shallow breathing	
*Kussmaul breathing	Vomiting	
Flushed skin	Plasma CO_2 increased	
*Dehydration		
*Vomiting		
*Urine excretion		
Hypoglycemia	Hyperglycemia	
Hunger	Thirst	
Sweating or cold and clammy	*Vomiting	
	Flushed face	
Nervousness	*Dry skin	
Faintness	*Restlessness	
Tremor	*Deep, rapid respiration (Kussmaul breathing)	
Slurred speech		
Tachycardia		
Headache	*Rapid pulse	
Irrational behavior	*Drowsiness	
	Low pulse pressure	
Alert the physician of the patient's physical changes and laboratory changes.		The physician will adjust his plan of care for the patient according to the physical changes and laboratory findings.

*Mrs. Thompson's symptoms of hyperglycemia and metabolic acidosis.

196

The nursing interventions in caring for Mrs. Thompson while she is receiving replacement therapy would be:

( ) a. Prepare the intravenous fluids sodium bicarbonate, potassium chloride, and insulin according to the physician's orders.

( ) b. Know the action, dosage, and side and adverse reactions to the parenteral and drug therapy.

( ) c. Observe for her adverse reactions from fluid and drug therapy.

( ) d. Make the changes in drug therapy as the nurse sees fit to do.

( ) e. Alert the physician of Mrs. Thompson's physical changes and laboratory changes.

----

a. X   d. –
b. X   e. X
c. X

197

Check the following symptoms that the nurse should be observing while Mrs. Thompson is receiving drug and parenteral therapy.

( ) a. hyponatremia
( ) b. hypernatremia
( ) c. hypovolemia
( ) d. hypervolemia
( ) e. hypocalcemia
( ) f. hypercalcemia
( ) g. hypokalemia
( ) h. hyperkalemia
( ) i. metabolic acidosis
( ) j. metabolic alkalosis
( ) k. hypoglycemia
( ) l. hyperglycemia
( ) m. renal function

----

a. X   f. –   j. X
b. X   g. X   k. X
c. X   h. X   l. X
d. X   i. X   m. X
e. –

Review—Diabetic Acidosis

Q. 1
Increased glucose concentration results in osmotic diuresis since it draws
fluid from * _____ .

— — — — — — — — — — — — — — —

the body tissues

_____

Q. 2
Osmotic diuresis inhibits reabsorption of _____ .

— — — — — — — — — — — — — — —

sodium or water

_____

Q. 3
The ketonic anions, from fat catabolism, will increase at the expense of
the _____ ion, resulting in ketosis and metabolic

_____ .

— — — — — — — — — — — — — — —

bicarbonate
acidosis

_____

Q. 4
Polyuria can occur from acetonuria and _____ .

— — — — — — — — — — — — — — —

glycosuria

_____

Q. 5
With the diabetic acidotic individual, dehydration can result from
nausea and _____ , _____ ,
and/or _____ .

— — — — — — — — — — — — — — —

vomiting, Kussmaul breathing, polyuria

_____

Q. 6
There is a serum potassium loss due to vomiting and renal excretion, but the hemoconcentration can cause the serum potassium to show

_____ .

_ _ _ _ _ _ _ _ _ _ _ _ _ _ _ _ _

normal or high values

Q. 7
Bed rest is helpful for the ill diabetic individual since it reduces the need

for _____ .

_ _ _ _ _ _ _ _ _ _ _ _ _ _ _ _

metabolism (reduces the need for glucose)

Q. 8
An individual receiving treatment for diabetic acidosis should be observed closely for a _____ reaction.

_ _ _ _ _ _ _ _ _ _ _ _ _ _ _ _

hypoglycemic

Q. 9
Mrs. Thompson's frequent vomiting and dry mucosa were an indication

of _____ .

_ _ _ _ _ _ _ _ _ _ _ _ _ _ _

dehydration

Q. 10
If glycosuria were present, would Mrs. Thompson's urinary specific

gravity be high or low? _____ .

_ _ _ _ _ _ _ _ _ _ _ _ _ _ _ _

high

Q. 11

Mrs. Thompson received normal saline intravenously to replace the sodium and chloride losses, and ampuls of $NaHCO_3$ to elevate her _____

_____ .

– – – – – – – – – – – – – – – –

plasma CO_2

Q. 12

Since Mrs. Thompson is receiving intravenous fluids, sodium bicarbonate, potassium chloride, and insulin, the nurse should be observing for

hypokalemia, hyperkalemia, _____ , _____ ,

_____ , _____ ,

_____ , _____ ,

_____ , _____ , and _____ .

– – – – – – – – – – – – – – –

hypovolemia, hypervolemia, hyponatremia, hypernatremia, metabolic acidosis, metabolic alkalosis, hypoglycemia, hyperglycemia, and renal function

SUMMARY QUESTIONS

The questions in this summary will test your comprehensive ability and knowledge of the material found in Chapter 6. If the answer is unknown, refer to the section in the chapter for further understanding and clarification.

Gastrointestinal Surgery

1. Many individuals undergoing gastrointestinal surgery will have had a previous fluid and electrolyte imbalance along with concurrent losses. At what times would replacement of fluid and electrolytes be indicated for the surgical patient?

2. Explain the reasons for hydrating Mr. Drum before surgery and for the insertion of the Levine tube preoperatively.
3. Answer the following questions concerning the Levine tube.
   a. Why is irrigation necessary?
   b. What are two ways to check patency and ensure proper drainage?
   c. What is one adverse effect of irrigation?
   d. How would you check for patency without the use of water?
4. For what purpose did Mr. Drum receive intravenous fluids containing dextrose, saline, and lactated Ringer's solutions? Why was sodium bicarbonate added to his intravenous fluid?

Cirrhosis of the Liver

5. Liver disease is frequently associated with the retention of _____

   _____ and _____ caused by_____

   and _____.
   Then why does hyponatremia occur in liver disease? Give other contributing causes of hyponatremia.
6. Explain the reasons for potassium loss frequently associated with liver disease.
7. Explain the role each of the following physiological factors plays in the development of ascites.
   a. Portal obstruction and hypertension
   b. Increased capillary permeability
   c. Decreased plasma osmotic pressure
   d. Retention of sodium and water
8. What is a paracentesis and when is it indicated? Explain the effects of repeated paracenteses in regard to fluid and electrolytes.
9. What is frequently the after-effect of abdominal paracentesis and what complication can accompany the after-effect?
10. Mr. Moore received digitalis and diuretics. Explain the use of Digoxin and the reason he received this drug. Explain the effects of Furosemide (Lasix) and Spironolactone (Aldactone) and how they differ from each other. What should the nurse be observing when patients receive continuous doses of diuretics and digitalis?

Renal Failure and Peritoneal Dialysis

11. What is a renal failure? Differentiate between acute and chronic renal failure. Explain the oliguric phase and diuretic phase in acute renal failure. What are the causes of acute and chronic renal failure?

12. Explain how the following fluid and electrolyte changes are similar or different for acute and chronic renal failure: potassium, sodium, calcium, overhydration, and metabolic acidosis. What symptoms of imbalance should the nurse be observing?

13. What is peritoneal dialysis? What substances can be removed from the body by this procedure?

14. The fluid used for peritoneal dialysis resembles _____. Differentiate between the two strengths of dextrose dialysate solutions and indicate the uses for each solution.

15. Why is potassium only added to the dialyzing solutions and not always included in commercially prepared solutions?

16. What is meant by positive fluid balance and negative fluid balance? At what point should a positive fluid balance be reported to the

    physician? _____. And a negative fluid balance?

    _____.

17. Indicate whether the following admission results are high, low, or normal. If high or low, what might the imbalance indicate?
    a. Hemoglobin 6.1 grams and hematocrit 19%
    b. B.U.N. 78 mg/100 ml and creatinine 4 mg/100 ml
    c. Plasma CO_2 13 mEq/L
    d. Plasma Cl 92 mEq/L
    e. Plasma Na 121 mEq/L
    f. Plasma K 5.6 mEq/L
    g. Plasma Ca 8.8 mg/100 ml

18. What types of food were restricted from Mrs. Grady's diet? Give examples of what foods should be restricted.

Burns

19. Following a burn, there is a shift of fluid and electrolytes from the

    _____ to the _____ of the
    burned areas. As a result of the shift what changes will occur in
    water, protein, and electrolytes? When, following a burn, does the
    greatest fluid shift occur.

20. Why does hemoconcentration occur in early burns and not later
    (when treatment is instituted)?

21. What are the three degrees in burns and give their characteristics.
    What is the big hazard that occurs in severe burns?

22. Explain the Rule of Nines in regard to burns.

23. Explain Cope and Moore's theory on the management of severe
    burns. Explain the Brooke Army Hospital formula in calculating
    parenteral therapy.

24. What solutions are used for colloid and electrolyte replacements?

25. What fluid and electrolyte imbalance should the nurse be observing
    while caring for Mrs. Silver and/or any patient with burns?

Diabetic Acidosis

26. What effects does an increased glucose concentration have on the
    glomerular filtrate?

27. Failure to metabolize glucose will lead to fat and protein

    _____.

28. Give the reasons for dehydration in diabetic acidosis. What electro-
    lyte changes occur?

29. In the clinical management of diabetic acidosis, what deficiencies
    are restored in the first 24 hours? What effect does rest have on
    metabolism?

30. What symptoms of fluid and electrolyte imbalance should the nurse
    be observing when caring for Mrs. Thompson before and after treat-
    ment?

31. The oral fluids which are tolerated soon after nausea and vomiting are ginger ale, broth, and orange juice. Indicate which of the following items (electrolytes, water, and sugar) are found in broth and orange juice.

	Broth	Orange Juice
Calcium	( )	( )
Potassium	( )	( )
Sodium	( )	( )
Sugar	( )	( )
Water	( )	( )

References

Abbott Laboratories, *Fluid and Electrolytes,* North Chicago, Illinois, 1968, pp. 31–36 and 39–42.

Baxter Laboratories, Inc., *Fluid Therapy,* Morton Grove, Illinois, 1962, pp. 68–73.

Beilski, Mary T. and David W. Molander, "Laennec's Cirrhosis," *American Journal of Nursing,* LXV (August, 1965), pp. 82–86.

Beland, Irene L., *Clinical Nursing: Pathophysiological and Psychosocial Approaches,* New York: The Macmillan Co., 1965, pp. 1275–1345.

Carnes, H. E., ed., "Acute Renal Failure," *Therapeutic Notes,* LXXI (January, 1964), pp. 8–12 (Parke, Davis and Co.).

Carnes, H. E., ed., "Treatment of Burns," *Therapeutic Notes* (October, 1963), pp. 234–239 (Parke, Davis and Co.).

Collentine, George E., "How to Calculate Fluids for Burned Patients," *American Journal of Nursing,* LXII (March, 1962), pp. 77–79.

Downing, Shirley R. and F. Louise Watkins, "The Patient Having Peritoneal Dialysis," *American Journal of Nursing,* LXVI (July, 1966), pp. 1572–1577.

Garb, Solomon, *Laboratory Tests in Common Use,* 4th ed. New York: Springer Publishing Co., 1966.

Hopps, Howard, *Principles of Pathology,* 2nd ed., New York: Appleton-Century-Crofts, 1964, pp. 302–303.

Lakey, William H., "Renal Failure," *The Canadian Nurse,* LIX (September, 1963).

Lubash, Glenn D., "Acute Renal Failure," *Hospital Medicine,* I (April, 1965), pp. 14–19 (Wallace Laboratories).

Martin, Marguerite M., "The Unconscious Diabetic Patient," *American Journal of Nursing,* LXI (November, 1961).

Metheney, Norma M. and William D. Snively, *Nurses' Handbook of Fluid Balance,* Philadelphia, Pa.: J. B. Lippincott Co., 1967, pp. 147–148, 182–183, 184–192, 216–223.

Nardi, George L. and George D. Zuidema, eds., *Surgery,* 2nd ed., Boston, Mass.: Little, Brown and Co., 1965, pp. 156–157.

O'Neill, Mary, "Peritoneal Dialysis," *The Nursing Clinics of North America,* I (June, 1966), pp. 319–323.

Statland, Harry, *Fluid and Electrolytes in Practice,* 3rd ed., Philadelphia, Pa.: J. B. Lippincott Co., 1963, pp. 63–75, 254–263, 276–280, 281–284.

Taber's Cyclopedic Medical Dictionary, 10th ed., Philadelphia, Pa.: F. A. Davis Co., 1965.

Trusk, Carol W., "Hemodialysis for Acute Renal Failure," *American Journal of Nursing,* LXV (February, 1965), pp. 80–85.

Appendix

The appendices that follow are guides for nurses in dealing with some clinical situations on medical and surgical floors in a hospital setting. These guides should increase the ability of the nurse to quickly assess fluid and electrolyte needs and changes occurring with their patients, and to assess means to evaluate and identify patients' needs based on observation and other available information.

There are five clinical situations presented here for the nurse's assessment of fluid and electrolyte imbalance. The situations presented are commonly encountered in the clinical setting. Appendix 1 relates to patients receiving parenteral therapy while being hospitalized regardless of their clinical problem. Therefore, the nurse must assess fluid and electrolyte changes occurring in the patient while he is receiving parenteral therapy. The outline presented in Appendix 2 on surgery will enable the nurse to make the necessary assessments before and following surgery. The three common medical problems are covered in the last three appendices. Presented in Appendix 3 is the patient with congestive heart failure, which is frequently a complication of many medical problems, e.g., myocardial infarction (a heart attack), arteriosclerosis, and renal and pulmonary diseases. Appendix 4 relates to patients in either an acute or chronic renal failure. Presented in Appendix 5 is the diabetic patient in diabetic acidosis.

The general outline of information necessary for the assessment of patients' body fluid and electrolyte needs and changes is as follows:

a. The knowledge and understanding of the clinical situation involved.
b. The clinical condition of the patient.
c. The types of therapy being administered.
d. The possible observed reactions to the therapy.

The above outline may be used for clinical situations other than those presented in these appendices.

Appendix 1

A Guide to Nurse's Assessment of Parenteral Therapy

Knowledge and Understanding of the Solution:

Hypotonic
Isotonic
Hypertonic
Potassium added
Other electrolytes contained in the solution, i.e., Na, Cl
Rate of flow

Clinical Condition of the Patient:
Age
General physical status

Types of Therapy:

Hydration
Maintenance
Replacement

Possible Reactions to Therapy:

Overhydration
Renal function
Shock
Infiltration
Phlebitis or thrombophlebitis
Potassium, sodium, and chloride deficits, and/or excesses

Appendix 2

A Guide to Nurses' Assessment of Patients Having Surgery

Knowledge and Understanding of the Surgery:

If GI surgery, the nurse should observe for electrolyte imbalance. Certain electrolytes are more plentiful in the stomach than in the intestine. Note which electrolytes are concentrated in body fluids as presented in the table.

Body fluid	Na^+	K^+	Cl^-	HCO_3-	(In mEq/L)
Gastric juice	60.4	9.2*	84	0–14	and H^+
Small bowel	111.3*	4.6	104.2*	31*	

*Electrolytes which are highly concentrated in these areas.

Clinical Condition of the Patient:

Preoperatively
    Dehydration
    Renal function

Postoperatively
    Retention or excretion of body fluid
      Water intoxication
      Edema
      Dehydration
    Sweating, which would be nonmeasured fluid loss
    Hyperventilation might result in respiratory alkalosis
    Hypoventilation might result in respiratory acidosis
    Vomiting and gastric intubation might result in metabolic alkalosis
    Diarrhea and intestinal intubation might result in metabolic acidosis
    Shock as the result of hypovolemia

Types of Therapy:
    Hydration therapy
    Replacement and maintenance therapy
    Gastric or intestinal intubation

Possible Observed Reactions to Therapy:
    Overhydration (hypervolemia)
    Abnormal laboratory results
    Output: Urinary—Dehydration, edema, or water intoxication
           GI drainage—intubation
           Nonmeasured fluid losses—sweating
    Shock (hypovolemia)

A Guide to Nurses' Assessment of Patients in Congestive Heart Failure

Knowledge and Understanding of Congestive Heart Failure:

> Increased venous pressure and increased production of aldosterone result in sodium and water retention or edema

Clinical Condition of the Patient:

> Signs and symptoms observed:
> Shortness of breath (dyspnea) and cyanosis
> Ankle edema
> Jugular vein engorgement—distended neck vein
> Increased pulse rate

Types of Therapy—"The Three D's":

Diet:	2–4 Gm daily requirement
	Limit sodium intake
	Digitalis leaf, digoxin, or digitoxin
Digitalization:	Digitalis intoxication with hypokalemia
Diuretics:	Sulfamyl ⎫
	Thiazide ⎬ Cause sodium and
	Osmotic ⎪ potassium losses
	Mercuric ⎭

> Aldosterone antagonists—cause sodium loss but not potassium

Possible Observed Reactions to Therapy:

> Diuresis—Potassium depletion when receiving digitalis and thiazide diuretics
>
> Retention of body fluid—Sodium excess and/or refractory to diuretics

Appendix 4

A Guide to Nurses' Assessment of Patients in Renal Failure Having Peritoneal Dialysis

Knowledge and Understanding of Renal Failure and Dialysis:

    Urinary output less than 400 ml daily
    Dialysis lowers the by-products of protein metabolism

Clinical Condition of the Patient:

    Anuria for 24 hours or more
    Pitted ankle edema
    No bladder distention
    B.U.N. $\uparrow$, CO_2 $\downarrow$, K $\uparrow$

Types of Therapy:

    Peritoneal Dialysis

        1-1/2% or 7% peritoneal dialysis fluid or dialysate
        KCL — 4 mEq/L added to fluid
        Diuretics

Possible Observed Reactions to Therapy:

    Positive and negative fluid balance
    Abnormal laboratory results
    Respiratory embarrassment
    Shock—Hypovolemia
    Dehydration
    Overhydration
    Water intoxication

Appendix 5

A Guide to Nurses' Assessment of Patients with Diabetic Acidosis

Knowledge and Understanding of Diabetic Acidosis:

    The body's inability to metabolize sugar due to a deficiency in insulin
    Hyperglycemia—excess blood sugar
    Ketone bodies accumulate in the blood and urine

Clinical Condition of the Patient:

    Might have had a cold or virus
    Dehydration, vomiting, weakness
    Electrolyte imbalance: CO_2 — decrease
                        K   — increase or decrease
    Kussmaul breathing, deep, rapid breathing
    Extreme thirst
    Sugar and ketone bodies in the urine
    Hyperglycemia
    Serum ketone bodies or acetone

Types of Therapy:

    Hydration therapy
    Replacement and maintenance therapy
    Electrolyte (KCl, Na HCO_3) replacement
    Insulin

Possible Observed Reactions to Therapy:

    Overhydration
    Renal function
    Electrolyte imbalance: too much K, Na
      (most commonly   too little K, Na
            observed)   too much HCO_3 can cause metabolic alkalosis
    Hypoglycemia from rapid correction

Bibliography

I would specially like to cite the books and articles marked with an asterisk as being of special use to me in writing this book.

BOOKS

*Abbott Laboratories, *Fluid and Electrolytes*, North Chicago, Illinois, 1968.

Barbata, Jean C., *et al.*, *Medical-Surgical Nursing*, New York: G. P. Putnam's Sons, 1964.

Baxter Laboratories, Inc., *The Fundamentals of Body Water and Electrolytes*, Morton Grove, Illinois, 1967.

Beland, Irene L., *Clinical Nursing: Pathophysiological and Psychosocial Approaches*, New York: The Macmillan Co., 1965.

Best, Charles H. and Norman B. Taylor, *The Physiological Basis of Medical Practice*, 8th ed., Baltimore, Md.: Williams and Wilkins Co., 1966.

Bland, John W., *Clinical Metabolism of Body Water and Electrolytes*, Philadelphia, Pa.: W. B. Saunders Co., 1963.

Brunner, Lillian Sholtis, *et al.*, *Medical-Surgical Nursing*, Philadelphia, Pa.: J. B. Lippincott Co., 1964.

Cooper, Lenna, *et al.*, *Nutrition in Health and Disease*, 14th ed., Philadelphia, Pa.: J. B. Lippincott Co., 1963

Dutcher, Isabel E. and Sandra B. Fielo, *Water and Electrolytes*, New York: The Macmillan Co., 1967.

Flitter, Hessel H., *Introduction to Physics in Nursing*, St. Louis, Mo.: C. V. Mosby, 1962.

Gard, Solomon, *Laboratory Tests in Common Use*, 4th ed., New York: Springer Publishing Co., 1966.

*Guyton, Arthur C., *Function of the Human Body*, 2nd ed., Philadelphia, Pa.: W. B. Saunders Co., 1964.

Guyton, Arthur C., *Textbook of Medical Physiology*, 3rd ed., Philadelphia, Pa.: W. B. Saunders Co., 1966.

Harrison, T. R., ed., *Principles of Internal Medicine*, 4th ed., New York: McGraw-Hill Book Co., 1962.

Hopps, Howard, *Principles of Pathology*, 2nd ed., New York: Appleton-Century-Crofts, 1964.

*Jacob, Stanley W. and Clarice A. Francone, *Structure and Function in Man*, Philadelphia, Pa.: W. B. Saunders Co., 1965.

Kimber, Diana C., *et al., Anatomy and Physiology,* 14th ed., New York: The Macmillan Co., 1961.

Kleiner, Israel S. and James M. Orten, *Biochemistry,* 7th ed., St. Louis, Mo.: C. V. Mosby, 1966.

Krug, Elsie E., *Pharmacology in Nursing,* 9th ed., New York: Blakiston Division, McGraw-Hill Book Co., 1961.

Langley, L. L., *Outline of Physiology,* New York: Blakiston Division, McGraw-Hill Book Co., 1961.

*Metheney, Norma M. and William D. Snively, *Nurse's Handbook of Fluid Balance,* Philadelphia, Pa.: J. B. Lippincott Co., 1967.

Montag, Mildred and Ruth P. S. Swenson, *Fundamentals in Nursing Care,* 3rd ed., Philadelphia, Pa.: W. B. Saunders Co., 1959.

Musser, Ruth D. and Betty Lou Shubkagel, *Pharmacology and Therapeutics,* 3rd ed., New York: The Macmillan Co., 1965.

Nardi, George L. and George D. Zuidema, eds., *Surgery,* 2nd ed., Boston, Mass.: Little, Brown and Co., 1965.

Pace, Donald, *et al., Physiology and Anatomy,* New York: Thomas Y. Crowell Co., 1965.

Ruch, Theodore C. and Harry Patton, eds., *Physiology and Biophysics,* Philadelphia, Pa.: W. B. Saunders Co., 1965.

Shafer, Kathleen N., *et al., Medical-Surgical Nursing,* 3rd ed., St. Louis, Mo.: C. V. Mosby, 1964.

Smith, Dorothy W. and Claudia D. Gips, *Care of the Adult Patient,* Philadelphia, Pa.: J. B. Lippincott Co., 1963.

Snively, William D., *Sea Within,* Philadelphia, Pa.: J. B. Lippincott Co., 1960.

*Statland, Harry, *Fluid and Electrolytes in Practice,* 3rd ed., Philadelphia, Pa.: J. B. Lippincott Co., 1963.

Taber, Clarence W., *Taber's Cyclopedic Medical Dictionary,* 10th ed., Philadelphia, Pa.: F. A. Davis Co., 1965.

Turner, Dorothea, *Handbook of Diet Therapy,* 3rd ed., Illinois: University of Chicago Press, 1959.

Wilson, Eva D., *et al., Principles of Nutrition,* 2nd ed., New York: John Wiley and Sons, 1966.

Wohl, Michael G. and Robert S. Goodhart, *Modern Nutrition in Health and Disease,* 3rd ed., Philadelphia, Pa.: Lea and Febiger, 1964.

PERIODICALS

Bielski, Mary T. and David W. Molander, "Laennec's Cirrhosis," *American Journal of Nursing,* LXV (August, 1965), pp. 82–86.

"Blood Volume," *Pitoclinic,* XI (March, 1964), pp. 5–10 (Ames Co.).

*Burgess, Richard E., "Fluids and Electrolytes," *American Journal of Nursing,* LXV (October, 1965), pp. 90–95.

Carnes, H. E., ed., "Acute Renal Failure," *Therapeutic Notes,* LXXI (January, 1964), pp. 8–12 (Parke Davis and Co.).

Carnes, H. E., ed., "Treatment of Burns," *Therapeutic Notes,* LXX (October, 1963), pp. 234–239 (Parke, Davis and Co.).

Collentine, George E., "How to Calculate Fluids for Burned Patients," *American Journal of Nursing,* LXII (March, 1962), pp. 77–79.

Crowell, Caleb E., ed., "Programmed Instruction–Potassium Imbalance," *American Journal of Nursing,* LXVII (February, 1967), pp. 343–366.

DeVeber, George A., "Fluid and Electrolyte Problems in the Post-operative Period," *The Nursing Clinics of North America,* I (June, 1966), pp. 275–283.

Downing, Shiley R. and F. Louise Watkins, "The Patient Having Peritoneal Dialysis," *American Journal of Nursing,* LXVI (July, 1966), pp. 1572–1577.

"Fluid and Ion Regulation," *Hospital Focus* (March 1, 1963), pp. 1–4 (Knoll Pharmaceutical Co.).

Grollman, Arthur, "Diuretics," *American Journal of Nursing,* LXV (January, 1965), pp. 84–89.

Jenkinson, Vivien M., "Congestive Heart Failure," *Basic Medical-Surgical Nursing,* Dubuque, Iowa: William C. Brown Co. (1966), pp. 166–171.

Lakey, William H., "Renal Failure," *The Canadian Nurse,* LIX (September, 1963).

Lubash, Glenn D., "Acute Renal Failure," *Hospital Medicine,* I (April, 1965), pp. 14–19 (Wallace Laboratories).

Martin, Marguerite M., "The Unconscious Diabetic Patient," *American Journal of Nursing,* LXI (November, 1961).

Morris, Dona G., "The Patient in Cardiogenic Shock," *Cardiovascular Nursing* (July–August, 1969).

O'Neill, Mary, "Peritoneal Dialysis," *The Nursing Clinics of North America,* I (June, 1966), pp. 309–323.

"Shock," *Hospital Focus* (October 1, 1962), pp. 1–6 (Knoll Pharmaceutical Co.).

Simeone, F. A., "Shock: It's Nature and Treatment," *American Journal of Nursing,* LXVI (June, 1966), pp. 1386–1394.

Strickland, William M., "Replacement Therapy in Traumatic Shock," *Seminar Report,* VI (Spring, 1961), pp. 2–7 (Merck, Sharp and Dohme).

Trusk, Carol W., "Hemodialysis for Acute Renal Failure," *American Journal of Nursing,* LXV (February, 1965), pp. 80–85.

Glossary

abdomen — the portion of the body lying between the chest and the pelvis.

acid — any substance which is sour in taste and will neutralize a basic substance.

acid metabolites — see metabolites.

ACTH — abbreviation for adrenocorticotropic hormone. A hormone secreted by the hypophysis or pituitary gland. It stimulates the adrenal cortex to secrete cortisone.

albumin — simple protein. It is the main protein from the blood.
  serum — Main function is to maintain the colloid osmotic pressure of the blood.

alimentary — pertaining to nutrition; the alimentary tract is a digestive tube from the mouth to the anus.

alkaline — any substance that can neutralize an acid and when combined with an acid will form a salt.

alveolus — air sac or cell of the lung.

anesthetic — an agent causing an insensibility to pain or touch.

anion — a negative charged ion.

anorexia — a loss of appetite.

anoxia — oxygen deficiency.

anuria — a complete urinary suppression.

aortic arch — the arch of the aorta soon after it leaves the heart.

aphasia — a loss of the power to speak.

arrhythmia — irregular heart rhythm.

arterioles — minute arteries leading into a capillary.

arteriosclerosis — pertaining to thickening, hardening, and loss of elasticity of the walls of the blood vessels.

artery — a vessel carrying blood from the heart to the tissue.

ascites — an accumulation of serous fluid in the peritoneal cavity.

atrium — a chamber. In the heart it is the upper chamber of each half of the heart.

azotemia — an excessive quantity of nitrogenous waste products in the blood.

biliary — pertaining to or conveying bile.

blood — intravascular fluid composed of red and white blood cells and platelets.

bradycardia — a slow heart beat.

bronchiectasis — dilatation of a bronchus.

B.U.N. — abbreviation for blood urea nitrogen. Urea is a by-product of protein metabolism.

capillary — a minute blood vessel connecting the smallest arteries (arterioles) with the smallest veins (venules).

carbonic anhydrase inhibitor — an agent used as a diuretic which inhibits the enzyme, carbonic anhydrase.

carotid sinus — a dilated area at the bifurcation of the carotid artery which is richly supplied with sensory nerve endings of the sinus branch of the vagus nerve.

cation — a positive charged ion.

cerebrospinal fluid — fluid found and circulating through the brain and spinal cord.

cirrhosis — a chronic disease of the liver characterized by degenerative changes in the liver cells.

colloid osmotic pressure — pressure exerted by nondiffusible substances.

contraindication — nonindicated form of therapy.

conversion table —

1 kilogram (Kg) = 2.2 pounds (lbs)

1 gram = 1000 milligrams (mg) or 15 grains (gr)

1 liter (L) = 1 quart or 1000 milliliter (ml)

1 cubic centimeter (cc) = 1 milliliter (ml)

1 drop (gtt) = 1 minum (m)

qh = every hour

X = times

cortisone — a hormone secreted by the adrenal cortex.

crystalloids — diffusible substances dissolved in solution and which will pass through a selectively permeable membrane.

CVP — abbreviation for Central Venous Pressure. It is the venous pressure in the vena cava or the heart's right atrium.

cyanosis — a bluish or grayish discoloration of the skin due to a lack of oxygen in the hemoglobin of the blood.

deficit — a lack of.

dermis — the true skin layer.

dextrose — a simple sugar, also known as glucose.

diabetes acidosis — an excessive production of ketone bodies (acid) due to a lack of insulin and inability to utilize carbohydrates.

diabetes mellitus — a disorder of carbohydrate metabolism due to an inadequate production or utilization of insulin.

dialysate — an isotonic solution used in dialysis and which has similar electrolyte contents to plasma or Ringer's solution, with the exception of potassium.

diffusion — the movement of each molecule along its own pathway irrespective of all other molecules; going in various directions.

dissociation — a separation.

diuresis — an abnormal increase in urine excretion.

diuretics — drugs used to increase the secretion of urine.

duodenal — pertaining to the duodenum, which is the first part of the small intestine.

dyspnea — a labored or difficult breathing.

edema — an abnormal retention of fluid in the interstitial spaces.
    pitted — depression in the edematous tissue.
    pulmonary — fluid throughout the lung tissue.

e.g. — for example.

electrolyte — a substance which when in solution will conduct an electric current.

endothelium — flat cells that line the blood and lymphatic vessels.

enzyme — a catalyst, capable of inducing chemical changes in other substances.

epidermis — an outer layer of the skin.

erythrocytes — red blood cells.

excess — too much.

excretion — an elimination of waste products from the body.

excretory — pertaining to excretion.

extra — outside of.

febrile — pertaining to a fever.

flatus — gas in the alimentary tract.

generic name — reflects the chemical family to which a drug belongs. The name never changes.

globulin — a group of simple proteins.

glomerulus — a capillary loop enclosed within the Bowman's capsule of the kidney.

glucose — formed from carbohydrates during digestion and frequently called dextrose.

glycogen — a stored form of sugar in the liver or muscle which can be converted into glucose.

gtts — abbreviation for drop.

hematocrit — the volume of red blood cells or erythrocytes in a given volume of blood.

hemodilution — an increase in the volume of blood plasma due to a lack of red blood cells or an excess of intravascular fluid.

hemoglobin — conjugated protein consisting of iron-containing pigment in the erythrocyte.

hemolysis — the destruction of red blood cells with the liberation of hemoglobin.

hepatic coma — liver failure.

hernia — a protrusion of an organ through the wall of a cavity.
    inguinal — protrusion of the intestine at the inguinal opening.

homeostasis — uniformity or stability. State of equilibrium of the internal environment.

hormone — a chemical substance originating in an organ or gland, which travels through the blood and is capable of increasing body activity or secretion.
    ADH — abbreviation for antidiuretic hormone; a hormone to lessen urine secretion.

hydrostatic pressure — pressure of fluids at equilibrium.

hyperbaric oxygenation — oxygen under pressure carried in the plasma.

hypercalcemia — a high serum calcium.

hyperchloremia — a high serum chloride.

hyperglycemia — an increase in the blood sugar.

hyperkalemia — a high serum potassium.

hypernatremia — a high serum sodium.

hypertension — high blood pressure.
    essential — a high blood pressure which develops in the absence of kidney disease. It is also called primary hypertension.

hypertonic — a higher solute concentration than plasma.

hyperventilation — increased breathing or respiration.

hypervolemia — an increase in blood volume.

hypocalcemia — a low serum calcium.

hypochloremia — a low serum chloride.

hypokalemia — a low serum potassium.

hyponatremia — a low serum sodium.

hypophysis — the pituitary gland.

hypotension — a low blood pressure.

hypotonic — a lower solute concentration than plasma.

hypovolemia — a decrease in blood volume.

i.e. — that is.

incarcerated — constricted as an irreducible hernia.

infusion — an injection of a solution directly into the vein.

insensible perspiration — water loss by diffusion through the skin.

insulin — a hormone secreted by the beta cells of the islets of Langer-
hans found in the pancreas. It is important in the oxidation and
utilization of blood sugar (glucose).

inter — between.

intervention — action.

intra — within.

ion — a particle carrying either a positive or negative charge.

ionization — separation into ions.

ionizing — separating into ions.

isotonic — same solute concentration as plasma.

ketone bodies — oxidation of fatty acids.

Kussmaul breathing — hyperactive, abnormally vigorous breathing.

lassitude — weariness.

Lidocaine — also known as xylocaine. A drug used as a surface anesthetic. It also can be used to treat ventricular arrhythmias.

lymph — an alkaline fluid. It is similar to plasma except its protein contents is lower.

lymphatic system — the conveyance of lymph from the tissues to the
blood.

malaise — uneasiness, ill feeling.

medulla — the central portion of an organ; e.g., adrenal gland.

membrane — a layer of tissue that covers a surface or organ, or separates
a space.

mercurial diuretic — a drug which affects the proximal tubules of the
kidneys by inhibiting reabsorption of sodium.

metabolism — the physical and chemical changes involved in the utilization of particular substances.

metabolites — the by-products of cellular metabolism or catabolism.

milliequivalent — the chemical activity of elements.

milligram — measures the weight of ions.
milliosmol — 1/1000th of an osmol. It involves the osmotic activity of a solution.
molar — 1 gram molecular weight of a substance.
myocardium — the muscle of the heart.

narcotic — a drug that depresses the central nervous system, relieves pain, and might produce sleep.
necrosis — destroyed tissue.
nephritis — inflammation of the kidney.
nephrosis — degenerative changes in the kidney.
neuromuscular — pertaining to the nerve and muscle.

oliguria — a diminished amount of urine.
osmol — a unit of osmotic pressure.
osmosis — the passage of a solvent through a partition that separates solutions of different strengths.
osmotic pressure — the pressure or force which develops when two solutions are of different concentrations and are separated by a selectively permeable membrane.
oxygenation — the combination of oxygen in tissues, blood and such.
oxyhemoglobin — hemoglobin carrying oxygen.

pallor or pallid — pale.
paracentesis — the surgical puncture of a cavity, e.g., abdomen.
parathyroid — an endocrine gland secreting the hormone, parathormone, which regulates calcium and phosphorus metabolism.
parenteral therapy — introduction of fluids into the body by means other than the alimentary tract.
patency — the state of being opened.
perfusion — passing of fluid through body space.
pericardial sac — fibroserous sac enclosing the heart.
pericarditis — inflammation of the pericardium or the membrane which encloses the heart.
peristalsis — wavelike movement occurring with hollow tubes such as the intestine for the movement of contents.
peritoneal cavity — a lining covering the abdominal organs with the exclusion of the kidneys.
permeability — capability of fluids and/or substances, e.g., ions, to diffuse through a human membrane.
  selectively permeable membrane — refers to the human membrane.
  semipermeable membrane — refers to artificial membranes.

phlebitis — inflammation of the vein.

physiological — pertaining to body function.

pitted edema — see edema.

plasma — intravascular fluid composed of water, ions, and colloid. Plasma is frequently referred to as serum.

pleural cavity — the space between the two pleuras.

pleurisy — inflammation of the pleura or the membranes which enclose the lung.

polyionic — many ions or ionic changes.

polyuria — an excessive amount or discharge of urine.

porosity — the state of being porous.

portal circulation — circulation of the blood through the liver.

postoperative — following an operation.

prednisone — synthetic hormonal drug resembling cortisone.

pressoreceptors — sensory nerve ending in the aorta and carotid sinus, which when stimulated will cause a change in the blood pressure.

pressure gradient — the difference in pressure which makes the fluid flow.

protein — nitrogenous compounds essential to all living organisms.
   serum — relates to albumin and globulin.
   plasma — relates to albumin, globulin, and fibrinogen.

pruritis — itching.

psychogenic polydipsia — psychological effect of drinking excessive amounts of water.

pulmonary edema — see edema.

pulse pressure — the difference between the systolic pressure and the diastolic pressure.

râles — pertaining to rattle. It is the sound heard in the chest due to the passage of air through the bronchi which contain secretions or fluid.

rationale — the reason.

reabsorption — the act of absorbing again an excreted substance.

retention — retaining or holding back in the body.

sclerosis — hardening of an organ or tissue.

selectively permeable membrane — see permeability.

semipermeable membrane — see permeability.

sensible perspiration — the loss of water on the skin due to sweat gland activity.

serous cavity — a cavity lined by a serous membrane.

serum — consists of plasma minus the fibrinogen. It is the same as plasma except that after coagulation of blood, the fibrinogen is removed. Serum is frequently referred to as plasma.

sign — an objective evidence of an abnormal nature in the body.

solute — a substance dissolved in a solution.

solvent — a liquid with a substance in solution.

specific gravity — a weight of a substance, e.g., urine. Water has a specific gravity of 1.000. The specific gravity of urine is higher.

steroid — an organic compound. It is frequently referred to as an adrenal cortex hormone.

stress — effect of a harmful condition or disease(s) affecting the body.

sympathetic nervous system — a part of the autonomic nervous system. It can act in an emergency.

symptom — subjective indication.

tachycardia — a fast heart beat.

tetany — a nervous affection characterized by tonic spasms of muscles.

thrombophlebitis — inflammation of a vein with a thrombus or a blood clot.

trade name — the name of the drug given by its manufacturer.

transudation — the passage of fluid through the pores of a membrane.

trauma — an injury.

urea — the final product of protein metabolism which is normally excreted by the kidneys.

uremia — a toxic condition due to the retention of nitrogenous substances (protein by-products) which cannot be excreted by the kidneys.

vasoconstriction — constriction of blood vessels.

vasodilatation — dilatation of blood vessels.

vasomotor — pertaining to the nerves having a muscular contraction or relaxation control of the blood vessel walls.

vasopressors — drugs given to contract muscles of the blood vessel walls to increase the blood pressure.

vein — a vessel carrying unoxygenated blood to the heart.

ventilation — the circulation of air.

 pulmonary — the inspiration and expiration of air from the lungs.

ventricles — the lower chambers of the heart.

venules — minute veins moving from capillaries.

vertigo — dizziness.

Index